Mapping Society

Mapping Society

*The Spatial Dimensions
of Social Cartography*

Laura Vaughan

First published in 2018 by
UCL Press
University College London
Gower Street
London WC1E 6BT

Available to download free: www.ucl.ac.uk/ucl-press

Text © Laura Vaughan, 2018
Images © Copyright holders named in captions, 2018

Laura Vaughan has asserted her right under the Copyright, Designs and Patents Act 1988 to be identified as author of this work.

A CIP catalogue record for this book is available from The British Library.

This book is published under a Creative Commons Attribution Non-commercial Non-derivative 4.0 International license (CC BY-NC-ND 4.0). This license allows you to share, copy, distribute and transmit the work for personal and non-commercial use providing author and publisher attribution is clearly stated. Attribution should include the following information:

Vaughan, L. 2018. *Mapping Society: The Spatial Dimensions of Social Cartography*. London, UCL Press. https://doi.org/10.14324/111. 9781787353053

Further details about Creative Commons licenses are available at http://creativecommons.org/licenses/

ISBN: 978–1–78735–307–7 (Hbk.)
ISBN: 978–1–78735–306–0 (Pbk.)
ISBN: 978–1–78735–305–3 (PDF)
ISBN: 978–1–78735–308–4 (epub)
ISBN: 978–1–78735–309–1 (mobi)
ISBN: 978–1–78735–310–7 (html)
DOI: https://doi.org/10.14324/111.9781787353053

To Neil and Daniel from 'U'

Foreword

The city is a human construct within which the different facets of the social condition evolve and exist. In this scholarly and absorbing volume, Professor Laura Vaughan deconstructs 'The City' in order to explore its constituents. In the context of this book 'The City' is not bound by one spatial or temporal location. Rather the volume focuses on the urban environment per se as the place under the microscope. Vaughan uses history as one of her tools to analyse the development of urban society and its social problems. One of the core themes of this impressive work is the way in which, over time, urban expansion and the parallel growth of a socio-economic infrastructure resulted in the creation of a disadvantaged and dispossessed sector of 'The City'. However, it is the way in which negative aspects of 'The City' – namely disease and poverty, crime and disorder and segregation as the result of differences of nationality, race and religion – have been highlighted by means of mapping that forms the spine of this book. As Professor of Urban Form and Society at UCL and an expert in the discipline of space syntax, it is not surprising that Vaughan has used her discipline's theories and methodologies as her main tool. Yet though this is a sophisticated and complex mode of research and analysis its use in this book is by no means inaccessible. The author ensures that her field of study, when applied, is eminently comprehensible, and is complemented by historical, geographical and sociological associations.

By the close of the nineteenth century London was a city without equal, its population numbering some 6.5 million. But the growth of global dominance and power was accompanied by the damaging features referred to above. And, as some gloried in the glittering aspects of the capital city others, particularly Charles Booth, worried about the 'problem of problems – poverty'. As the chapters in this book highlight, while Booth was not the first to use mapping to illustrate the negative characteristics of a city, his mapping of the levels of poverty in certain areas of London – though, as Vaughan tells us, very much ignored by later social studies – was at the time received as a 'most remarkable and important contribution'.[1] Booth's first map of poverty, published in 1889, shocked society. It was, according to *The Guardian*, 'a physical chart of sorrow, suffering and crime'[2] which, *The Times* recorded, 'drew the curtain behind which East London had been hidden'.[3] The map provided an instantaneous coloured depiction of the darker side of London; the rare instances of affluence signalled in yellow and red descending to the all-too-frequent presence of darkest blue or black poverty. Ten years later Booth updated and added to his original map of poverty and in an absorbing chapter Vaughan reproduces these maps, providing detailed textual analysis which highlights the ways in which urban space had expanded and been violated by the development of the capital's transport system, new roads and railways frequently separating the poor and underclass from the rest of society.

Though London may have been the leading city of the latter half of the nineteenth century, Vaughan ensures that her readers are introduced to the way in which mapping has been used from the early nineteenth century by social enquirers in a range of disciplines in other major metropolises both at home and abroad. The multi-ethnic mix and variance of wage rates in Chicago were published in mapping form (and accompanied by explanatory essays) in the *Hull-House Maps and Papers* in 1895. These were edited by the social reformer Jane Adams, who had visited and been inspired by Toynbee Hall, the settlement in London's East End founded in 1884. Adams acknowledged her debt to Booth, stating that she had used the same colour coding as he had in his 'wage maps' – as she referred to his maps of poverty. Though the recording of ethnic difference was not on Booth's agenda,[4] it was on that of W.E.B. Du Bois, an African American who mapped the location of 'Negroes' in 1890s Philadelphia. Once again visual cartographic immediacy adds impact and import to the location of late nineteenth century African-Americans in a leading North American city. 1890s Chicago and Philadelphia cartography indeed informed where and how ethnic minorities in both cities were located and segregated. Some 120 years later what Booth, Du Bois, Hull-House and others achieved by arduous street tramping research, the ethnographic, social and religious make-up of cities such as London, can now be visually presented as the result of the fusing of up-to-the minute data and advanced computer technology.

The impressive collection of maps which illustrate the book's chapters range temporally from one created by Leonardo da Vinci, of the town plan of Imola, in 1502, to two produced in 2015, one highlighting levels of multiple deprivation in the Limehouse area of London and the other identifying the excessive walking distance from the peripheral *banlieues* to the nearest Paris Metro stations. These graphic illustrations of disadvantage provide an instantly visible insight into just some of the urban social problems that are extant in the early twenty-first century.

Though mapping is at the core of this book, the reader cannot ignore the array of photographs that Vaughan has used to supplement the mapped information, images which reinforce the reality of the deprivation and hardship that was, and in some instances still is, the lot of the poorer members of society. Indeed, one of the important lessons of this book is that social research needs to combine textual analysis with maps and images. In this way the composition – spatial syntax – of the urban environment can be explored through a selection of lenses which enable both the instant recognition of 'the problem of problems' through maps and images and its implications as they emerge in accompanying documented data; this combination *can* – and should – lead the way to the amelioration of urban social problems.

Whilst this volume was written from the viewpoint of an expert in urban form and society, its appeal goes far beyond that discipline. Academically it is multi-disciplinary, but its reach should not stop there: the quality and variety of the maps and photographs, together with its accessible text, makes it a book for all those concerned to understand the way in which the growth of the urban environment determined the way in which people lived – and still live – in 'The City'.

Anne Kershen
Queen Mary University of London

Notes

1. *Daily News*, 6 April 1889.
2. *The Guardian*, 17 April 1889.
3. *The Times*, 15 April 1889.
4. The one exception (though it was not commissioned by Charles Booth) was the mapping of the Jewish community of East London which appeared in the Russell and Lewis book *The Jew in London*. This was published in 1901 and is reproduced and discussed in this book.

Preface

Sitting in a seminar room at UCL in 1992 while studying for my Master's in Advanced Architectural Studies (nowadays the MSc Space Syntax), it was seeing Charles Booth's maps of poverty for the first time that sparked my interest in what has become a career-long project of unpacking the social complexity of cities. The period of the early 1990s was a time when Bill Hillier, the founder of the field of space syntax, was developing his conception of the city as a 'movement economy', a theory which viewed the city as organising different land uses and economic classes within the same area by separating them marginally, with only a turning or two separating the rich and poor. Using the Booth map as a source to identify an urban phenomenon allowed Hillier to hypothesise that the way in which nineteenth-century London organised class spatially was the outcome of a dynamic process that first shaped flows of movement through a city's streets and then arranged land uses to follow the spatial logic of a city. By identifying a phenomenon of 'marginal separation by linear integration', Hillier was using the historical map not only as a source of information on how cities worked in the past, but also as source of inspiration for building a broad theory of how cities work in general. My work builds on this tradition and on the incredible body of scholarship constructed by Bill Hillier, Julienne Hanson, Alan Penn and myriad colleagues within the space syntax community, from which I have spun out my own specific combination of urban design, geographical and historical readings of urban society.

A few months after my first introduction to the Booth map I was browsing in the Hebrew and Jewish Studies section of UCL library and came across a fragile book from 1901, with an even more fragile map inside: the map of Jewish East London, 1899. Looking at the way the map, with its shadings of blue from light to dark, was used to accentuate the density of Jewish immigrant settlement in the area, immediately struck me as showing some fundamental spatial regularities beyond simply being a ghetto – as it was known then. This was the start of many years of research into the complexity of socio-spatial segregation and one direct outcome was an intensive study of the Booth maps of poverty in London, in which I used space syntax methods to consider if there is a relationship between social and spatial exclusion.

In fact, this book reflects two decades of enquiry into the spatial nature of society, with a specific focus on the detailed patterning of social phenomena as these are laid out in historical maps. The importance of a spatial analysis of historical data is not to be underestimated. Going beyond placing the data on the map to a deeper analysis of the geographical patterning of the data allows the researcher to pose a variety of questions: regarding the spatial character of the urban setting, regarding whether social data of a single type have spatial characteristics in common, and – in general – to control for spatial effects when analysing social patterns.

For me, the Booth maps have become the quintessential starting point when exploring the relationship between the spatial organisation of cities and how societies take shape over time. I have written about the spatial-temporal evolution of cities in my earlier book, *Suburban Urbanities*. This current publication effectively goes back to the origins of my research, starting with my most fundamental subject of interest: how the spatial configuration of cities shapes social patterns and, specifically, urban social problems.

This book does so by taking maps of social statistics and developing a close reading of the maps themselves as well as the context within which they were created. A side product of this inquiry has been the discovery of the extent to which social cartography is frequently used not only as a tool for communicating information on patterns of settlement, but also for other purposes: for propaganda, to collate evidence or to support scientific argumentation. The use of social maps as an analytical device is less prevalent and this book will show how a reading of the spatial patterns captured by such maps can reveal some fundamental rules about how cities work according to a specifically spatial logic of society.

Among the most famous examples of social cartography are John Snow's maps of cholera in London and, from later in the nineteenth century, Charles Booth's maps of poverty. Other, less well-known maps, such as the Russell and Lewis map of Jewish East London, stemmed from a desire to capture the apparent disorder of urban areas and to solve what continue to be viewed as urban problems: slums or ghettos.

Writing about cities as socio-spatial artefacts demands a theoretical approach which sees the physical city as a measurable artefact. By considering social space as one of the many variables which make up the complex patterns of urban cityscapes, it is possible to gain deeper insights into the significance of the spatial patterns of the data that are captured in social cartography. The complexity of these patterns requires an ability to describe and analyse them beyond simple descriptions of patterns on the ground. The use of *space syntax* theories and methods, which have at their heart the concept of spatial *configuration* – namely a consideration of how each space in the city relates to other spaces, results in a mathematical reading of social space that has, since its inception in the 1970s, led to hundreds of studies and thousands of practitioners worldwide. The introduction to this book will explain these theories in greater depth and the reader will observe, throughout the book, points where space syntax theories are used to illuminate the cartographic representations of society and so to illustrate how spatial configuration is a fundamental (and oft-overlooked) aspect of social cartography. The application of space syntax theories in this book is primarily as a way of considering socio-spatial patterns systematically, but the appendix additionally provides a description of the fundamentals of space syntax theory as well as a worked-through example of analysis. Moreover, where relevant, the endnotes indicate where other detailed scientific expositions of space syntax analysis can be obtained.

Ultimately this book's ambition is to demonstrate how an interdisciplinary reading of social maps can provide a richer understanding of how society and urban spatial systems interact with each other. Thus, phenomena such as segregation can only be fully understood once we take account of a wide variety of factors, including economic, political, social as well as spatial context – and all this in addition to the changes that cities and their inhabitants undergo over time.

Acknowledgements

I open by thanking Anne Kershen, founder of the Centre for the Study of Migration, Queen Mary University of London, whose work as a historian and expert in migration has been an inspiration to me since we first met over two decades ago. Her constant support as a mentor over the years has shown an intellectual (and practical) generosity I can only aspire to in my own academic practice.

I am also grateful to Lara Speicher from UCL Press, who has encouraged this project ever since I first mentioned to her my interest in historic maps. Thanks must also go to the team at the press; especially Jaimee Biggins and Alison Fox.

While most of the research reported here has been done alongside other academic activities, I would like to acknowledge the sabbatical funding from the Architecture Research Fund of the Bartlett School of Architecture, UCL in 2017, as well as research funding received from the UK Engineering and Physical Sciences Research Council, the latter of which supported my analysis of several of the maps in this volume.[1]

I would also like to thank the many people, mostly anonymous, who have donated maps to various repositories around the world, as well as the curators of some important map collections, without whom this study would have taken a significantly longer time. A special mention is due to the David Rumsey Collection and the University of Chicago Map Collection, as well as to Cornell University Library's Persuasive Maps Collection whose vast repositories were invaluable for my research and whose (mostly anonymous) archivists and curators provided some excellent leads for background reading on the maps.

I am grateful to the following people and institutions who granted me permission to use their own map copies, in most cases responding within a number of hours to my request: Emmanuel Eliot and the Geoconfluences project; Stephen Ferguson from Princeton University Library; the Science and Society Picture Library, the Science Museum; Ralph R. Frerichs, the UCLA Department of Epidemiology, School of Public Health; Sam Brown from the University of Chicago Map Collection; Jenny Schrader for information on the Blind Alleys of Washington; Chris Mullen for providing (and scanning for me) his copy of the Rowntree map of York; the University of Pennsylvania archives and Amy Hillier for access to the Du Bois map; Oliver O'Brien, Consumer Data Research Centre (CDRC); Alex Werner, curator, Museum of London; Tinho da Cruz, librarian, Department of Geography and Planning, University of Liverpool, who brought the Hume map of Liverpool to my attention and provided me with a copy of it; Edward Denison and the Asmara Heritage Project; and Jonathan Potter Ltd., whose responsiveness, as a commercial business, is one of many examples of the generosity I have enjoyed throughout this project.

Note

1. The funding in question was an EPSRC PhD studentship, which supported research into patterns of Jewish settlement in nineteenth-century Manchester and Leeds, and an EPSRC 'First Grant' (no. GR/S26163/01), which ran from 2003–5, funding research into the spatial form of poverty in nineteenth-century London using the Charles Booth maps of poverty. This was titled 'Space and exclusion: the relationship between physical segregation and economic marginalisation in the urban environment'.

Contents

List of figures	*xvi*
1. Mapping the spatial logic of society	1
2. Disease, health and housing	24
3. Charles Booth and the mapping of poverty	61
4. Poverty mapping after Charles Booth	93
5. Nationalities, race and religion	129
6. Crime and disorder	168
7. Conclusions	205
Appendix: The spatial syntax of society	*223*
References	*230*
Index	*243*

List of figures

Figure 1.1	Leonardo da Vinci's town plan of Imola, c. 1502.	3
Figure 1.2	Plate I from Seaman's *An Inquiry into the Cause of the Prevalence of the Yellow Fever in New-York*, 10 March 1797.	5
Figure 1.3	Gustave Doré, *Over London, by Rail*, 1872.	9
Figure 1.4	*Plan of the environs of Venice showing the lagunes*, &c, 1838.	14
Figure 1.5	Detail of the plan of Venice, showing the position of the Ghetto north of the Grand Canal.	15
Figure 2.1	Detail of Plate II from Seaman's *An Inquiry into the Cause of the Prevalence of the Yellow Fever in New-York*, 10 March 1797.	26
Figure 2.2	*Sanitary Map of the Town of Leeds*, 1842.	28
Figure 2.3	Detail showing Kirkgate area of *Sanitary Map of the Town of Leeds*, 1842.	30
Figure 2.4	Back-to-backs on Nelson Street, Leeds, c.1901.	31
Figure 2.5	*Cholera in Rouen*, 1832.	32
Figure 2.6	Detail of *Cholera in Rouen*, 1832.	33
Figure 2.7	Space syntax analysis of the Soho area c.1890 and Ordnance Survey map of London of the same period.	34
Figure 2.8	J. Snow, *Street Map of Soho, around Golden Square, Illustrating Incidences of Cholera Deaths during the Period of the Cholera Epidemic*, 1853.	36
Figure 2.9	Detail of Fig. 2.8, J. Snow, *Street Map of Soho*, 1853.	36
Figure 2.10	Photographs of tenement housing styles in New York: 'old-law' railroad-type housing, 'old-law' dumb-bell-type housing, and 'new-law' tenement housing.	41
Figure 2.11	'Lodgers in a Crowded Bayard Street Tenement – "Five Cents a Spot"', Lower East Side, New York, 1889 Jacob Riis, 1889.	42

Figure 2.12	The Tenement House Committee maps, 1894.	43
Figure 2.13	The Lung Block, New York, 1903.	44
Figure 2.14	*Tuberculosis in a Congested District in Chicago*, 1 January 1906 to 1 January 1908.	46
Figure 2.15	Detail of *Tuberculosis in a Congested District in Chicago*.	47
Figure 2.16	*The Blind Alley of Washington: Seclusion Breeding Crime and Disease*, 1911.	48
Figure 2.17	*Children's Automobile 'Accidents' in Detroit*, 1970.	51
Figure 2.18	Rue Estienne, Paris, *c*.1853.	54
Figure 2.19	Rooftop *chambres de bonne* in Paris, 2008.	55
Figure 3.1	Rev. Abraham Hume's *Map of Liverpool, Ecclesiastical and Social*, 1854.	62
Figure 3.2	Detail of Rev. Abraham Hume's *Map of Liverpool, Ecclesiastical and Social*.	64
Figure 3.3	Key to Rev. Abraham Hume's *Map of Liverpool, Ecclesiastical and Social*.	65
Figure 3.4	Thomas Hawksley's *Plan Shewing the Arrangement of 'Back to Back' Houses*, 1844.	65
Figure 3.5	Charles Booth's *Descriptive Map of London Poverty*, 1889, sheets 1–4 compiled into a single image	72
Figure 3.6	'George H. Duckworth's Notebook: Police and Publicans District 14 [West Hackney and South East Islington], District 15 [South West Islington], District 16 [Highbury, Stoke Newington, Stamford Hill]', 1897, p. 199.	74
Figure 3.7	Detail of *Descriptive Map of London Poverty*, 1889.	75
Figure 3.8	Charles Booth's *Map Descriptive of London Poverty*, 1898–9, sheet 6.	78
Figure 3.9	Detail of Charles Booth's *Map Descriptive of London Poverty*, 1898–9, sheet 6.	79
Figure 3.10	a) Detail of *Descriptive Map of London Poverty*, 1889 and b) detail of the same area on the *Map Descriptive of London Poverty*, 1898–9, sheet 6.	82
Figure 3.11	Abingdon House, Boundary Estate, Old Nichol Street, 2016.	83
Figure 3.12	Detail of *Descriptive Map of London Poverty*, 1889.	84
Figure 4.1	*Map of York Showing the Position of the Licensed Houses*, 1901.	95
Figure 4.2	Detail of *Map of York Showing the Position of the Licensed Houses*, 1901.	98

Figure 4.3	Hull-House wages maps 1–4, 1895.	100
Figure 4.4	Detail of wage map no. 4, Hull-House wages maps.	102
Figure 4.5	Detail of wage map no. 2, Hull-House wages maps.	104
Figure 4.6	A typical back yard in the slum district of Chicago, c.1905.	106
Figure 4.7	*The Seventh Ward of Philadelphia*, 1896.	111
Figure 4.8	Detail, *The Seventh Ward of Philadelphia*.	112
Figure 4.9	*City and Rural Population*, c.1890.	114
Figure 4.10	View from a rooftop of the corner of Ocean Street and Masters Street, Stepney, London 1937.	116
Figure 4.11	*The New Survey of London Life and Labour*; key to map.	118
Figure 4.12	Detail of *Map Descriptive of London Poverty*, 1898–9, sheet 1 with inset of same area from *New Survey of London Life and Labour*, 1929–31.	119
Figure 4.13	Detail of *Map Descriptive of London Poverty*, 1898–9, sheet 1 with inset of same area from *New Survey of London Life and Labour*, 1929–31.	120
Figure 4.14	Limehouse area of London showing Index of Multiple Deprivation (2015) ratings.	122
Figure 5.1	*Official Map of Chinatown*, San Francisco, 1885.	134
Figure 5.2	The Street of Gamblers (Ross Alley), San Francisco, California (1898).	135
Figure 5.3	An opium den, Chinatown, San Francisco, California, (c.1900).	135
Figure 5.4	Detail of Fig. 5.1, *Official Map of Chinatown*, San Francisco, 1885.	137
Figure 5.5	Detail of Fig. 5.1, *Official Map of Chinatown*, San Francisco, 1885.	138
Figure 5.6	Hull-House nationalities maps 1–4, Chicago, 1895.	140
Figure 5.7	Detail of Hull-House nationalities maps 1–4, Chicago, 1895.	142
Figure 5.8	*A Map of Newark With Areas Where Different Nationalities Predominate*.	144
Figure 5.9	*Jewish East London*, 1899.	145
Figure 5.10	Key to map of *Jewish East London*.	147
Figure 5.11	Detail from Charles Booth, *Maps Descriptive of London Poverty*, 1898–9.	148
Figure 5.12	Detail from map of *Jewish East London,* 1899.	149
Figure 5.13	Chevrah Shass Synagogue, Old Montague Street, East London c.1950.	152

Figure 5.14	Detail of Goad Plan, sheet 322, 1899, showing location of Old Montague Street Synagogue (Chevrah Shass).	153
Figure 5.15	*Racial Zoning Map of the City of Asmara*, 1916.	154
Figure 5.16	Redline map of Miami, Florida *c.*1933.	157
Figure 5.17	Key to Miami redline map.	158
Figure 6.1	*Portsoken Ward with its Divisions into Parishes*, 1772.	172
Figure 6.2	*Nineteenth Precinct, First Ward, Chicago*, 1893.	176
Figure 6.3	*The Modern Plague of London*, *c.*1884.	180
Figure 6.4	Detail of *Modern Plague of London*.	181
Figure 6.5	*Map Showing Places of Religious Worship, Public Elementary Schools, and Houses Licensed for the Sale of Intoxicating Drinks*, London, 1900.	182
Figure 6.6	Key to *Map Showing Places of Religious Worship*.	183
Figure 6.7	Detail of *Map Showing Places of Religious Worship*.	183
Figure 6.8	*Liquordom in New York City*, 1883.	187
Figure 6.9	Saloon Map of New York City, 1888	189
Figure 6.10	Detail of *The Temperance Movement* overlaid on space syntax analysis of Manhattan *c.*1891.	190
Figure 6.11	Detail from *A scene in the ghetto, Hester Street, New York*, *c.*1902.	191
Figure 6.12	*Chart 2 – Urban Areas*, illustrating the growth of cities, from Burgess, 1925.	193
Figure 6.13	*Map no. VII* showing places of residence of alleged male offenders in Cook County jail, 1920.	195
Figure 6.14	Detail of *Map no. VII*.	196
Figure 6.15	*Chicago's Gangland*, 1927.	197
Figure 6.16	Detail of *Chicago's Gangland*.	199
Figure 7.1	Paris: wealth disparities, 2015.	212
Figure 7.2	15-minute walking distance from a train station, Paris, 2015.	213
Figure 7.3	The Venice Ghetto and its houses of windows, 2016.	217
Figure a	Space is not a background to activity, but an intrinsic aspect of it.	224
Figure b	Axial line map of East End district of London, overlaid on Booth map 1889.	226
Figure c	Detail of *Descriptive Map of London Poverty* 1889, overlaid with space syntax analysis.	227
Figure d	Detail of *Descriptive Map of London Poverty* 1889.	228
Figure e	Detail of space syntax analysis.	228

1
Mapping the spatial logic of society

Spatial order is one of the most striking means by which we recognise the existence of the cultural differences between one social formation and another, that is, differences in the ways in which members of those societies live out and reproduce their social existence.[1]

The origins of social cartography

Despite its shifting meaning, mapping remains 'a way of representing the world', a visible image of, if not the world, then an aspect of that world.[2] This book is about a very specific type of mapping, social cartography – namely the creation of maps whose purpose is to represent specific aspects of society at a given time and place. Histories of maps abound; where this book differs is in its aim to convey how social maps can also be records of social enquiry in relation to the role of urban configuration in shaping social patterns over time. It will demonstrate how a better understanding of the relationship between society and space can shed light on fundamental urban phenomena that normally tend to be seen purely as by-products of social structures. By bringing spatial analysis into the foreground, it emphasises the power of space in shaping society over time. In addition, by tracing a long century's worth of social cartography from the 1790s onwards, this book aims to provide an overview of a sub-set of maps, demonstrating not only their graphic power in conveying data, but also the way in which each of them expresses a single point in urban history, marking key points in the evolution of urban social space.

Coincidental with the concern with urban ills throughout the latter half of the nineteenth century were developments in cartography that enabled map makers to convey complex population statistics illustratively by presenting them on a map. It is this context of urban upheaval caused by industrialisation that explains why nineteenth-century Britain became one source (but not the only, as we will see) of the phenomenon of the social reformer as urban investigator, replete with maps and statistical data for surveying and tabulating the 'uncharted' territory of these mushrooming urban settlements. As social environments these rapidly growing cities were as alien to the understanding and taste of classically educated elites as those of imperial acquisitions in India and Africa were to the mass of the British population. The founder of the Salvation Army William Booth made this explicit by comparing England's 'darkest' slums to those of Africa: 'The foul and fetid breath of our slums is almost as poisonous as that of the African swamp.'[3] At about the same time William Booth was writing this, his contemporary, the social reformer Charles Booth, was producing his extraordinary street maps which revealed to the world the extreme contrasts of wealth and poverty that existed cheek by jowl in London. The development of mapping technology and practices, from the Ordnance Survey to temperance organisations, from municipal boundary commissions to Goad fire insurance plans, all emerged in the context of the nineteenth-century enthusiasm to map the uncharted 'urban interior' of cities as well as to bring scientific rigour to analysing and solving the many ills that had befallen cities.

The scope of this book is the century following the earliest experimentations in putting statistics on maps: starting with a map of yellow fever from late-eighteenth-century New York, through to maps of contemporary cities across the world. In its earliest incarnation, the social map was concerned with epidemic disease, particularly cholera. It associated disease with problems of urban poverty and the need for sanitary improvement of cities followed logically. Later in the nineteenth century the perceived problem of mass migration to the growing cities led to the application of segregation as a political device to separate disparate populations. Emerging from usage within laboratories (where segregation simply meant separation of materials or chemicals), 'segregation' took on a weightier meaning with its evolution as a synonym for quarantines against disease contagion and – subsequently – for racial separation.

While public health had been one of the most important drivers of early nineteenth-century mapping of disease, the re-emergence of the bubonic plague in Hong Kong in 1894 led to fresh calls to separate European and native populations in territories across Africa and Asia, while in countries such as the United States and Australia that had received mass migration from those continents there was an increasing interest in putting up immigration barriers. Whether these decisions were driven purely by concerns with public health or by deeper problems of racism, the outcome was the same: the creation of sharp spatial barriers to prevent social mixing.

The use of mapping as a tool of social investigation reached its peak with the emergence of a science of social investigation in the 1880s. It is unlikely to be a coincidence to find that the largest number of maps found in the archival research undertaken for this book date from the late nineteenth- to early twentieth-century period, which coincides with the publication and mass dissemination of Charles Booth's revolutionary methods of social cartography. Earlier urban maps go back centuries, such as Da Vinci's 1502 mapping of the defensive potential of Cesare Borgia's conquests (see Figure 1.1): this example shows precise cartographic rendition of the town of Imola, using for the first time a true planar view instead of the perspectival view more common at the time.[4]

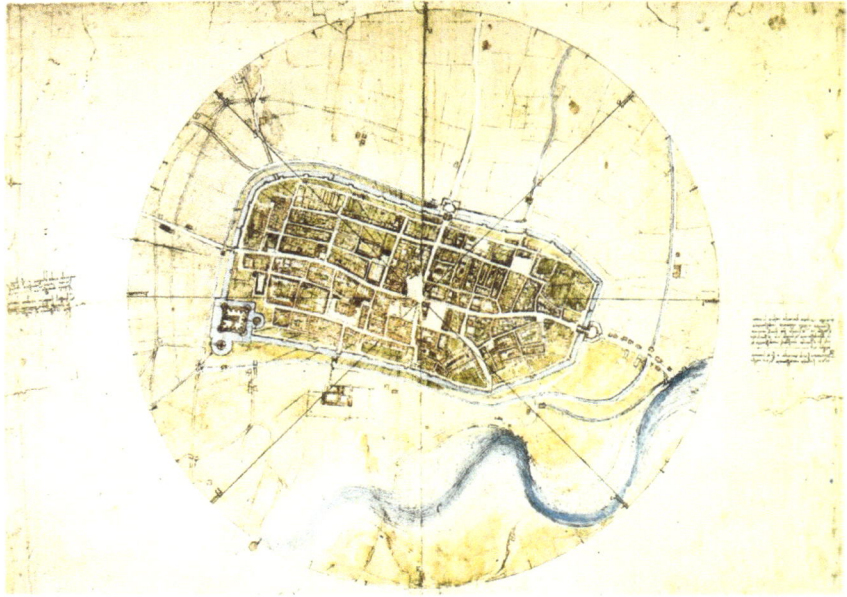

Figure 1.1 Leonardo da Vinci's town plan of Imola, c. 1502.

Pencil, chalk, pen and wash on paper, 440 x 602 mm Museo Vinciano, Vinci.
Public domain via Wikimedia Commons.

Yet, in trying to pinpoint a logical scope for this book, I have chosen to start with the earliest instances of social cartography. This starting point is used to focus on the use of maps for social investigation, instead of purely as data records. The aim is to consider both the *intention* – is it a map whose purpose is social enquiry – as well as the *outcome* – does it capture data in sufficient detail that we can learn from it something about how urban society worked at the time? This book is particularly concerned with maps which record data at the scale of buildings or streets, as they give a detailed rendition of the lived space of the social city at the time.

The map of Imola constitutes another important point in history for this spatially focused book. While it used an arbitrarily defined centre, in this instance a crossroads, Da Vinci was the first to bring an objective measurement system into his depiction of the town plan, in contrast with the geometric forms, such as squares or circles, that one might see in earlier maps. Thus, as Robin Evans states, geometry relinquished its direct hold on the form it described.[5] Ultimately, maps were shifting from having a symbolic power to having a descriptive (arguably scientific) power.

Another reason to open this survey of social cartography at the start of the nineteenth century is that most histories of cartography mark the start of the century with the first attempts to plot diseases on maps in dot form. One of the earliest examples of these is Valentine Seaman's maps of yellow fever, published 1800, which used dots and circles to show individual occurrences of the disease in the waterfront areas of New York (see his Plate 1 in Figure 1.2, detail).[6] Subsequently Alexandre Parent-Duchâtelet's extensive data tabulation, time series, and mapping of prostitutes in Paris posthumously published in 1836 marked an important development in cartography by using a choropleth map, which shades geographical areas according to data averages.[7]

As we will see in this book's chapter on health and disease, dot maps shifted from being purely a graphic device to becoming a way in which to render statistics graphically and then to hypothesise on the actual causes of disease. From maps concerned with the literal infection of the urban body, other forms of social ill, such as crime and poverty, also became subject to mapping analysis.

Social cartography became more prevalent throughout the nineteenth century as cities grew, alongside which a growing concern with the moral, as well as the physical, health of the nation became more central to public debate. A series of developments in cartography as well as

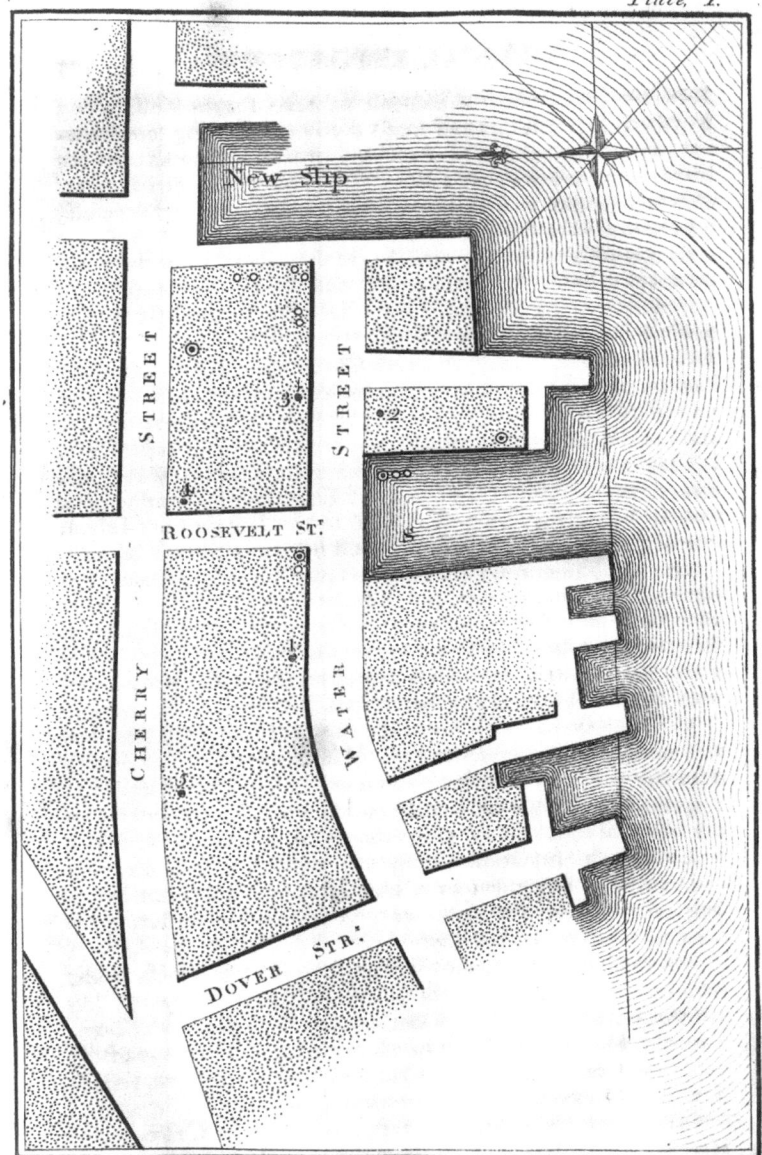

Figure 1.2 Plate I from Seaman's *An Inquiry into the Cause of the Prevalence of the Yellow Fever in New-York*, 10 March 1797.

Medical Repository, 1 (1800, 2nd edition): 303–323 [Rare Books Collection]. Image copyright Princeton University.

the development of lithographic printing (which enabled commercial printing of maps in colour) assisted this growth in social mapping, such that the 20-year period leading up to 1835 has been termed a 'golden age' in geographic cartography, in which the fundamental methods for representing population numbers, distribution, density and movements came into being.[8] Alongside this, the growth of social statistics (censuses, but also surveys and other forms of population study), and rapid developments in statistical graphics from 1786 onwards (first the line graph and bar chart, then the pie chart and circle graph in 1801) and in statistical maps (with the first use of continuous shading to show geographical distributions in 1826) together made for an ideal setting in which social maps would become an increasingly popular form of communication throughout the latter half of the nineteenth century.[9]

In writing this book an attempt has been made to seek out as wide a range of social maps as possible, though it is clear from earlier scholars that despite the fact that religious diversity, for example, has been a feature of some cities in Asia for millennia and an interest in surveying and mapping society goes back to antiquity, the drive to map this diversity, and to consider it as a phenomenon worth recording, ameliorating or reforming, seems to be a particularly western issue.[10] Thus the book's geographical scope is centred on London, with examples from across Europe and (in particular) the United States. The focus on London is not accidental: it was at the forefront of cartographic innovation and it was amongst the first cities in modern times to experience mass urbanisation, and hence, to have a need to manage a diverse, densely crowded population.

Visual rhetoric

Maps claim to be objective, to record facts on the ground, but even the most sophisticated cartographic productions can contain errors. It was only in 2012 that Sandy Island in the South Pacific was discovered actually not to exist, despite having featured on maps, both printed and electronic, for over a century (it was apparently the result of a mistaken record by whalers who encountered a reef in its purported location). This argument is well-rehearsed: aside from errors, the positioning of maps (as well as the choice of where they end) can affect the way in which the world is viewed. The sixth-century Madaba Map shows Jerusalem at the centre of the world, while the sixteenth-century Mercator projection – still used as the basis for world mapping today – gives the readers of the standard map of the world the impression that land masses close to the poles are much larger than they are in reality.

Maps are social constructions, whose integrity as scientific objects is limited to how precise they are when taking account of their scale and similar measurable parameters. As soon as decisions start to be made on selection of data and the way in which those data are to be presented on maps, the social and political context in which they were created will start to influence how they are read. Once this fact is recognised, one can get beyond the traditional criticism of maps and start to consider what they are in reality: objects laden with meaning, which reflect the context of their creation. Yet maps continue to be incredibly useful for capturing data as well as for providing a starting point for analysing those data statistically.

Geographers often view maps as an outcome of the research process. In contrast historians tend to consider them as representations of a given social and cultural milieu; others, such as social or political scientists, will use maps simply to illustrate an argument, or to provide a snapshot of a situation at a given time. Their importance as sources for information on the spatial structure of society is understood to a much lesser degree and is sometimes avoided entirely, for fear of deterministic interpretations of how the built environment affects social outcomes. This is a critical gap that this book aims to fill. It aims to convey how cartographic records of society reveal much more than a social pattern on the ground. It will show how maps can also be records of social enquiry; specifically, in relation to the role of urban configurations in shaping social patterns, in some cases over time. It will demonstrate how a better understanding of the relationship between society and space can shed light on fundamental urban phenomena, such as segregation, which tend to be viewed purely as by-products of social structures.

Yet, it is important to recognise an essential problem with the study of social maps in rapidly changing populations, such as those found in large cities. They will portray the situation of a community as if in a fixed state (and as if the community itself is uniform in its constitution). The reality of course is that both maps and social data in general can impose 'a false appearance of stability on what were economically and culturally unstable cities and city-spaces'. Nor will they capture, as Richard Dennis has put it, 'the diurnal rhythms of society'; rather, they will record who was sleeping where on a specific night and 'not necessarily where they normally slept and not where, or with whom they spent their work or leisure time'.[11]

A similar challenge lies with the reliance on official maps as objective records of a society's situation at a given time. It is essential to take account of the social and political context within which a map is constructed, before reading it as a historical source. In the case of thematic maps, for example, not only is there the potential for data distortions in choosing where to place category intervals (say, between one class and the next), but other fundamental weaknesses stem from what is known as the ecological fallacy, namely the assumption that if two sets of data coincide on a map, that one is the cause of the other.[12] An example of this is found in the analysis by Malgaigne of the incidence of hiatus hernia in France, set against levels of consumption of olive oil. His map illustrates the incidence of hernias among the population of French departments by shading at six intervals the number of people per 'hernious' individual within each administrative unit, and then superimposing on it a line that represented the limits of olive cultivation.[13] Denis Cosgrove has written about the importance of Malgaigne's map in the history of cartography. By testing a hypothesis with a combination of statistical and cartographic methods, Malgaigne had created the first ever statistical map of disease (and indeed used it to refute the prevailing belief that differences in consumption of olive oil were the cause of a higher incidence of hernia amongst the Midi population).[14]

Another inherent weakness with social maps is what Cosgrove has termed the 'iconographic' weakness – the choice of colour to illustrate categories ('interval differences'). Notably, he states, the concept of shading itself can carry 'powerful moral connotations, especially in the past, 'when darkness and shadow implied ignorance and decay, both physical and moral'.[15] Examples of the use of darkness to hint at immorality abound. From Charles Booth's use of darker shades of blue for the poverty classes in his maps of poverty in London (blue itself being much cooler than the bright reds and golds used for the richer classes), to the map of 'Jewish East London' drawn up by Booth's associate, George Arkell, which used a similar visual rhetoric to plot the percentages of Jewish settlement in shades of blue for the majority streets and shades of red for the Jewish minority streets. The result was a map which showed the patches of majority Jewish immigrant settlement, which were in fact relatively small, seemingly much larger due to their being coloured in the deeper, darker shades of blue, while minority Jewish streets were coloured in the brighter shades of pink and red.

Both Miles Kimball and Pamela Gilbert refer to Booth's use of warm colours for wealth, through cooler colours to the darkest colours for poverty. Gilbert argues that 'it worked at such a sophisticated level of particularity that it became possible to trace the trajectory of the spatial development and movement of affluence over time'.[16] Interestingly, this recalls Newman's notes on the Arkell map of Jewish East London, which states how the map 'ossifies an essentially fluid situation'.[17] The growth of colour printing at this time was to effect a shift in the use of visual rhetoric in public campaigns regarding urban social ills, which prior to the late nineteenth century had been seen as a uniform mass of darkness.

Such rhetoric is also evident from illustrations from the time, such as Gustave Doré's engraving of the London slums, *Over London, by Rail* (see Figure 1.3), which shows the rear yards of the terraced houses deep in the shadow of the railway viaduct. This view was evidently familiar to

Figure 1.3. Gustave Doré, *Over London, by Rail*, 1872.
Illustration for Blanchard Jerrold's *London: A Pilgrimage* (1872).
Copyright Science and Society Picture Library, The Science Museum.

Charles Booth, who spoke of such a scene in his 1887 paper to the Royal Statistical Society:

> Those of us who have seen no more, have at least obtained a sort of bird's-eye view of such places from the window of a railway carriage, passing along some viaduct raised above the chimneys of two-storied London. Seen from a distance, the clothes lines are the most visible thing. Those who have not such outside accommodation must dry the clothes in the room in which they eat, and very likely also sleep . . . From the railway, there may be seen, also, small rough-roofed erections, interspersed with little glass houses. These represent hobbies, pursuits of leisure hours – plants, flowers, fowls, pigeons, and there is room to sit out, when the weather is fine enough, with friend and pipe and glass. All this goes when the workshop invades the back yard; and as to sanitation and health, I need hardly point out how essential is sufficient space behind each house.
>
> Worse again than the interleaving of small cottage property or the addition of workshops, is the solid backward extension, whether for business premises or as tenements, or as common lodging houses, of the buildings which front the street; and finally we have quarters in which house reaches back to house, and means of communication are opened through and through, for the convenience and safeguard of the inhabitants in case of pursuit by the police.[18]

In fact, as Gareth Stedman-Jones has written, poor districts had become by this time 'an immense terra incognita periodically mapped out by intrepid missionaries and explorers who catered to an insatiable middle-class demand for travellers' tales'.[19] Lurid newspaper articles on the East End used 'slum stereotypes and other formulaic motifs' to reinforce the colourful descriptions read by the masses, helped by the fact that most of their readers had never ventured into its streets.[20] On the other hand, sensational imagery used by those who visited the slums safely at a distance from their carriages was bolstered by more precise accounts from what might be termed as explorers, who roamed the streets on foot to get closer to the reality of life in and on the slum streets. These ranged from Henry Mayhew's newspaper articles published between 1849 and 1850 (and collected in *London Labour and the London Poor*), through Charles Dickens on his *Night Walks*, whose accounts of 'houselessness' helped shift Victorian consciences regarding the plight of the poor. Dickens was ultimately supplanted by late-century social investigators, who transformed the presentation of the city, 'just as the tales of the explorer had given way to the maps of the administrators of the Empire'.[21] Most famously, Charles Booth's investigations were of the ilk of explorers whose aim was to draw up a topographical map of their adventures. By the end of the century Victorian rationality had virtually won its battle against Victorian sentimentality and sensationalism, as Kevin Bulmer has pointed out: Booth's importance in the development of the social sciences is that instead of simply seeking methods for the amelioration of poverty, his aim was to enquire into its nature scientifically.[22]

In fact, the graphic impact of this period's social cartography was not only on how people viewed urban problems, but also how they conceptualised them. Jacob Riis, whose photographic records of poverty in New York were used to show 'how the other half lives', wrote that if the city's streets were coloured to show the location of its various nationalities, the map 'would show more stripes than on the skin of a zebra, and more colours than any rainbow . . . [which] would give the whole the appearance of an extraordinary crazy quilt'.[23]

In this context it is interesting to contrast Riis' vision of New York City as a 'crazy quilt' with the concept of the 'melting pot' depicted by Kurt Weill in his opera *Street Scene*. According to Richard Dennis, Weill describes a tenement building as containing a suite for each ethnic group – Jews, Italians, Germans, Irish, Swedes and a black janitor – 'encapsulating the problems and the potential of the city's "melting pot" in a single building'.[24] The spatial patterning hinted at in these descriptions is essentially the main impetus that has driven the writing of this book. Too many social maps are used simply as descriptive accounts of the spatial patterns of society. In its subsequent chapters this book will show how, whether a 'crazy quilt', a 'melting pot' or otherwise, the organisation of society in urban space follows systematic regularities in the built fabric of cities that can be measured, quantified and analysed to powerful effect.

Linguistic colour

David Sibley has written how the Greeks and the Romans saw themselves as standing at the centre of the civilised world, so that the farther away a group was from the imperial hub, the 'greater was its "vice"'.[25] Any civilisation that was inferior to the Greek or Roman culture was in effect deviating from the mean – or the norm – in its statistical as well as its physical sense. A map of prostitution, he argues, provides essential information about 'the social topography of the town. The basic principle of medieval regulation was to designate certain areas to prostitution, either inside or outside the walls, and limit vice strictly to them.' This was a form of 'social hygiene', locating prostitution in poor districts, 'often close to the river' or beyond the city walls.[26]

By drawing boundaries around people other from themselves, European powers defined the separation of the centre from the periphery. This analysis recalls similar ideas put forward by Richard Sennett regarding the treatment of the Jews of Venice. In his history of Western civilisation, *Flesh and Stone*, Sennett writes of the 'fear of touching' that led the Christian community of Venice to seek to isolate its foreign inhabitants, as if they were 'isolating a disease that had infected the community . . . with corrupting bodily vices'. Sennett writes about how the Jewish body was especially regarded as unclean, thus the enforced isolation of the Jewish inhabitants of the Ghetto of Venice (named after its former purpose, a cannon foundry, or *ghèto* in Venetian Italian),[27] was a way to purify the Christian city.[28] The Jewish religion itself was perceived to be a contaminating influence and in the case of the holy city of Rome, the city's ghetto served to isolate non-Catholics from the remainder of the city's inhabitants.

At the time of the creation of its Ghetto, Venice was a globally strategic city, situated at a junction of trade routes that had turned it into a trade hub containing people of many different nationalities. While all minority groups had restrictions of one sort or another, the Jewish inhabitants of the city were placed under night-time curfew within the walled, moated island ghetto, with additional restrictions placed on their trading and social contact.

In contrast with the spatial isolation of the city's Jewish population, there are contemporaneous accounts which show that the Venice Ghetto's inhabitants used the city's network strategically to carry out day-to-day activities,[29]

> I went to the Ghetto where the Jews dwell as in a suburb by themselves being invited by a Jew of my acquaintance to see a circumcision. I passed by the Piazza Judea where their seraglio begins for

being inviron'd with walls they are lock'd up every night. In this place remains yet part of a stately fabric, which my Jew told me had been a palace of theirs for the ambassador of their nation when their country was subject to the Romans.[30]

At the same time as restricting their movement, the Venetian Ghetto and its counterparts elsewhere in Europe provided bodily security for Jews against religious persecution.[31] There are indeed some scholars who claim that the ghetto was a way for the authorities to create a spatial and temporal compromise between the economic contribution of the minority and the wish to exclude a perceived 'contaminating presence'.[32] The result was that certain categories of Jews were permitted to venture out of the ghetto after dark; physicians, for example, were allowed out at night 'to participate in scientific meetings or cure the sick'.[33] In fact, despite the ghetto acting as an exclusionary device, there were quite high rates of commercial and cultural interchange between the inhabitants of ghettos and wider society.[34]

This complex use of space belies the normal approach to interpreting segregated social space, which tends to focus on the residential location of a minority group, overlooking their opportunities for movement across the city, throughout the day and the week. Spatial scientists are starting to meet this challenge by distinguishing between the more extreme forms of geographical isolation and simple clustering of minority populations. In one interesting experiment, the physicists Volchenkov and Blanchard have analysed how easy it might have been in the past to get around Venice's canal network, creating a mathematical model of an imaginary set of gondoliers, with starting points distributed around the city's canals. Each gondolier is set on a random journey to see where they might end up.[35] The results of the experiment were that the most important canals, the Grand and the Giudecca, were used the most, being the most connected, and that the most isolated canal was that next to the Venetian Ghetto. This finding seems conclusive, showing that spatial separation corresponds to the Venetian desire to isolate the city's Jewish population.

However, Sophia Psarra's analysis of the city's historic network, which uses space syntax network measures to analyse the canal network on its own and then linked up with the street network, provides additional detail on the Venice Ghetto's spatial segregation, as it takes account of different degrees of accessibility, both local and city-wide. Psarra explains how Venice evolved over many years from relatively isolated

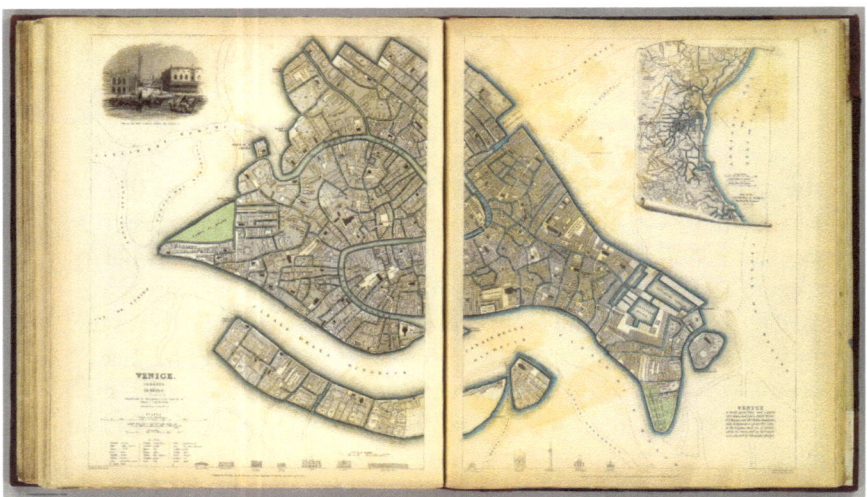

Figure 1.4 *Plan of the environs of Venice showing the lagunes, &c.*

Drawn by W.B. Clarke. Engraved by B.R. Davies, 16 George Street, Euston Sq. Published by the Society for the Diffusion of Useful Knowledge, 59 Lincoln's Inn Fields, 15 July 1838. (London: Chapman & Hall, 1844).

Image copyright 2000, Cartography Associates.

communities formed around a church and a square, situated in an archipelago of islands (and indeed the Ghetto was one of the most isolated of these islands, situated at the city edge). Over the centuries, as the islands were stitched together by canals and by land reclamation, streets and bridges, a much more integrated system was created to make wider connections across the city.[36] The Venice Ghetto, while not being very well connected to the city, was nevertheless quite accessible locally (see Figures 1.4 and 1.5, which show the city's street and canal network as it was at a later period, in 1838).[37]

Indeed, not only are there accounts of Jewish people buying and selling in the Rialto, but Donatella Calabi also describes how some worked in Christian-owned businesses or performed 'with their music and dance companies in the homes of Venetian aristocrats'.[38] Connections outside of the ghetto were farther reaching still, through international trade as well as the periodic waves of migration from distant lands as other European countries changed their policies towards their Jewish inhabitants.

Despite the complexity of the subject, many nineteenth-century texts continued to use powerfully negative imagery to describe the poor,

Figure 1.5 Detail of the plan of Venice, showing the position of the Ghetto north of the Grand Canal.

Image copyright 2000, Cartography Associates.

foreigners and other people viewed as marginal to society in emotive, sensual terms – emphasising their moral contagiousness. David Sibley has described how a distinction was made between the 'pure bourgeois and the defiled proletarian' in mid-nineteenth-century Paris as part of an effort to 'deodorize utopian city space' during the reshaping of the city under Baron Haussmann.[39] This type of distinction between people of differing social status is confirmed by Ryan Powell, who writes that the contemporary city uses space to socially segregate people in more subtle but equally powerful ways, showing that the placement of gypsy camps in marginal areas at the edges of cities results in gypsies living in 'marginal, yet ambivalent, spaces'.[40]

Minority groups are typically not the only people to be marginalised spatially. So too are the poorest classes. Despite the slum clearances that took place in London throughout the latter half of the nineteenth century, the increasing numbers of people moving into the city from the countryside, coupled with a lack of organised city planning, led to the formation of poverty areas constituted by an 'almost endless intricacy of courts and yards crossing each other . . . like a rabbit warren'.[41] Robin Evans has commented that the campaigns to clear the slums were due to their being viewed in the public eye at the time to be breeding grounds for indecency '. . . as if the homes had been one great block of stone eaten by slugs into innumerable small chambers and connecting passages'.[42] Overcrowding was linked with immorality, while poverty was associated linguistically with the animal (in this instance, the insect) world, suggesting the poor to be non-human in their behaviour. At the same time, likening the city's morphology – its physical form and layout – to a rabbit warren shows how the city was itself viewed by people as a source of the immorality of its inhabitants.[43]

No language, especially when used to describe marginality, is entirely neutral. When Engels was roaming the streets of London, visiting the slums, but also negotiating his way through the streets of 'The Great Town', he wrote: 'The very turmoil of the streets has something repulsive, something against which human nature rebels'.[44] This could be interpreted as a repulsion against the lack of care towards the city's poor, but equally, his words can be read as horror at the degradation of their behaviour as well. Similarly, numerous accounts of nineteenth-century slums show the negative perceptions associated with clusters of ethnic minorities. One example of this can be seen in a review of newspaper cuttings from the turn of the century. Here we see that despite the perceptions, or reality, of disease, degradation and danger, people ventured into the slums as if to view an exotic other world: 'In Sydney slumland is "discovered" by "municipal expeditions" which set forth to "explore" the "wilds." This sense of discovery is strengthened by analogies of foreign evangelisation and conquest . . . '[45]

Fear of the foreigner became especially acute after the arrival of immigrants into Britain and the United States in the 1880s and 1890s. Newspapers were full of outcries at the influx of 'foreigners', describing San Francisco's Chinatown, where a labourer was suspected of having died of bubonic plague, as a colony of 'disorder and sin', whose inhabitants, once the plague quarantine cordon had been lifted, started to 'swarm in and out'.⁴⁶ While many of the fears were unfounded, concerns over the impact of immigration on the American economy, labour market and culture still led to the Chinese Exclusion Act of 1882, a piece of federal legislation that halted the immigration of Chinese labourers to the United States for a decade.

This sense of foreign intrusion was strengthened by styling Chinatown a 'colony'. The term was widely used in American cities to characterise poverty districts as being foreign in their character. Looking at the map of San Francisco's Chinatown of that time (the first to be called that) we can see a compelling example of visual as well as linguistic rhetoric being used for a political purpose – to raise public concerns about the supposedly invasive population.⁴⁷ According to Carl Nightingale, 'the map helped give an official stamp to a widespread belief that the city's Chinese residents were the source of vice and disease and that their neighbourhood posed a threat to the city as a whole'.⁴⁸ This added negative factor of uncleanness, sin and squalor was common to public perceptions of ethnic enclaves, and will be examined in detail later in this book.

Mapping the complexity of society

Questions regarding the historiography of the nineteenth-century industrial city and of the large urban settlements characteristic of modernity demand analysis that can handle large amounts of data on urban societies while taking account of their spatial context (as well as their political, economic and other contexts). One recalls Durkheim's characterisation of this modernity in terms of its physical and moral *density*, in which he makes a connection between the form and pattern of relationships between people (what Durkheim refers to as 'social morphology').⁴⁹ Notably for this book's overarching purpose, to analyse the spatial logic of society, it is significant that although Durkheim and, latterly, Lévi-Strauss commented on the need to study the social structure of society by analysing its spatial arrangements, it is evident that not all spatial arrangements are a direct reflection of the societies (or communities) that they contain.⁵⁰

Space syntax methods provide a reading of urban environments as multi-scale entities that have evolved over time. Although cities have always been laid out with the intention of shaping how their inhabitants will interact and integrate, or indeed to stop them from doing so, they are in many cases emergent systems, evolving slowly as the result of piecemeal decisions made by individuals. This emergence may still involve planning interventions at key stages to shape patterns on the ground, but decisions made by people – such as where they wish to live – will also influence the way in which these complex systems operate. Similarly, the political, policy and social context of cities and urban societies will also influence the degree of autonomy individuals have over their lives. Evidently, the weaker one's position in society, the less power one has over where one can live; thus, the people who are most disadvantaged socially or economically will tend to end up in the least spatially advantageous places in the city, such as the areas cut off by the railways (see Chapter 2 section on Housing and Health). This claim is seemingly uncontentious, but for a spatial scientist to make the claim that cities have any agency in social outcomes most certainly is contentious.[51] An additional purpose of this book is therefore to provide the evidence for spatial configuration having an impact in shaping society, to show that segregation is a spatial as well as a social problem and to demonstrate the complexity even of its most extreme realisation: the ghetto.[52]

Space syntax research has shown that more complex societies, namely urban societies, are normally comprised of a *structured non-correspondence* between society and space, so that in locations such as 1950s Bethnal Green, London, or the West End of Boston, space is not structured to correspond to social groups. Rather it creates a social fabric in which the many different social groupings of urban complex society can take shape across the 'warp and weft' of socio-spatial relations, some of which require a strong correspondence to reinforce the relations within a group, others of which make do with weaker ties that are maintained across larger distances:

> Space may not be structured to correspond to social groups, and by implication to separate them, but on the contrary to create encounters among those whom the structures of social categories divide from each other. In other words, space can in principle also be structured, and play an important role in social relations by working against the tendency of social categorization to divide society into discrete groups. Space can also reassemble what society divides.[53]

The complexity in which cities evolve as part of a socio-spatial process reflects the city's three aspects. First, as a social structure, the city is composed of households, communities, kin relations and ethnic and social groups. Second, as an economy, the city is characterised by interlinked markets in goods, land and labour. Third, as a political structure, the city is composed of networks of institutions and interest groups. Cities are experienced by individuals instantaneously, and yet they grow and evolve over historic timescales, whilst many of their social and cultural properties are only recognisable as population-wide behaviour patterns. Yet there are complex interrelationships between the configuration of local streets and wider cross-city (as well as global) trajectories. This suggests that the layout of cities shapes the way in which urban spaces both acquire social meaning and have social consequences over time.[54] Attempting to unravel this complexity is undoubtedly a big challenge, but at the very least this book will show how social maps can be analysed to uncover the possible ways in which cities evolve to adapt to social needs (or indeed fail to do so, resulting in patterns of inequality).

John Snow's decision to map disease, like subsequent maps from later in the century – Charles Booth's maps of poverty and the Russell and Lewis map of Jewish East London – stemmed from a desire to capture the apparent disorder of urban areas and to solve what are still seen to be urban problems: rookeries, slums or ghettos. This book explores maps of social statistics to reveal how they served as a tool for communicating information on patterns of settlement, but also for other purposes: for propaganda, to collate evidence or to support scientific argumentation. Their use as an analytical device is less prevalent and this book aims to show how a reading of the spatial patterns captured by such maps can reveal some fundamental rules of how cities work according to a specifically spatial logic of society. Social maps reflect the urban conditions of a given society at a given time, yet they also provide an opportunity for analysing those conditions to see if there are any systemic explanations for the persistence of urban problems such as poverty, whether there are any underlying factors in the way in which minority groups settle in urban areas or how religious minorities take advantage of the spatial structure of the city. The city is not always a benign setting for emergent social systems to take shape; it can be used negatively, either through poor planning that creates unsafe streets or through a cynical use of the power of physical separation to remove unwanted groups from the city centre. By considering maps as researchable artefacts, we can put the spatiality of society at centre stage.

This argument is elaborated in the following chapters, starting with a chapter that considers the earliest form of social mapping: *disease* maps.

Subsequent chapters consider in turn the two main groups of social maps: maps of *poverty* – used both in the past and today to chart and thus diagnose this apparently insolvable urban problem; and maps of *nationalities, race and religion*, which tend to focus on the individual as a minority within society. The penultimate chapter considers what might be the essence of mapping of social malaise: maps of *crime and disorder*. The closing chapter discusses the future of social-spatial analysis in an era of online data availability and raises a word of warning to those considering the supposed democratising of data on issues such as crime as being necessarily a social advancement. The power maps have had in the past to effect positive change can be used also to ill effect, especially in such complex issues as segregation.

Notes

1. B. Hillier and J. Hanson, *The Social Logic of Space*, 1990 ed. (Cambridge: CUP, 1984), p. 27.
2. D. Cosgrove, *Geography and Vision: Seeing, Imagining and Representing the World* (London and New York: I.B.Tauris, 2012), p. 1.
3. General William Booth was founder of the Salvation Army. Quotation from General William Booth, *In Darkest England and the Way Out* (New York and London: Funk & Wagnal, 1890), p. 14.
4. L. Rombai, 'Cartography in the Central Italian States from 1480 to 1680,' in *The History of Cartography, Volume 3: Cartography in the European Renaissance, Part 1*, ed. D. Woodward (Chicago: University of Chicago Press, 2007), p. 935. More specifically, this was an exact survey-based plan – ichnographic, not iconographic – that is, showing buildings and public spaces to scale, rather than simply showing their appearance. M. Hebbert, 'Figure-Ground: History and Practice of a Planning Technique,' *Town Planning Review* 87, no. 6 (2016), p. 708.
5. R. Evans, *The Projective Cast: Architecture and Its Three Geometries* (London: Architectural Association, 1995), p. 45.
6. Plate II can be seen in Chapter 2, Figure 2.1.
7. See review of this and other cartographic innovations in M. Friendly and D.J. Denis, 'Milestones in the History of Thematic Cartography, Statistical Graphics, and Data Visualization,' http://datavis.ca/milestones/.
8. A.H. Robinson, 'The 1837 Maps of Henry Drury Harness,' *The Geographical Journal* 121, no. 4 (1955), p. 440.
9. Although this point can be debated, Martin Bulmer pinpoints the emergence of social investigation as dating from 1880. See M. Bulmer, K. Bales and K. Kish Sklar, eds, *The Social Survey in Historical Perspective, 1880–1940* (Cambridge: Cambridge University Press, 1991). For a review of this 'golden age' of statistical graphics, see M. Friendly, 'The Golden Age of Statistical Graphics,' *Statistical Sciences* 23, no. 4 (2008).
10. A.H. Robinson, *Early Thematic Mapping in the History of Cartography* (Chicago: University of Chicago Press, 1982).
11. R. Dennis, *Cities in Modernity: Representations and Productions of Metropolitan Space, 1840–1930* (Cambridge: Cambridge University Press, 2008), p. 69.
12. An early critique of this sort of fallacy appeared in W.S. Robinson, 'Ecological Correlations and the Behavior of Individuals,' *American Sociological Review* 15, no. 3 (1950).
13. Cosgrove, *Geography and Vision*, p. 164.
14. S. Jarcho, 'An Early Medicostatistical Map (Malgaigne, 1840),' *Bulletin of the New York Academy of Medicine* 50, no. 1 (1974).
15. Cosgrove, *Geography and Vision*, pp. 163–4.
16. M.A. Kimball, 'London through Rose-Colored Graphics: Visual Rhetoric and Information Graphic Design in Charles Booth's Maps of London Poverty,' *Journal of Technical Writing*

and *Communication* 36, no. 4 (2006); P.K. Gilbert, 'The Victorian Social Body and Urban Cartography,' in *Imagined Londons*, ed. P.K. Gilbert (Albany: State University of New York Press, 2002), p. 24.
17. A. Newman, *Jewish East London: The Russell and Lewis Map (Publisher's Note on Reproduction of Map from 'The Jew in London')* (London: The London Museum of Jewish Life, 1985), p. 4.
18. C. Booth, 'Condition and Occupations of the People of East London and Hackney, 1887,' *Journal of the Royal Statistical Society* 51, no. 2 (1888), pp. 284–5.
19. G. Stedman Jones, *Outcast London: A Study in the Relationship between Classes in Victorian Society*. Oxford: Peregrine Penguin Edition, 1984, p. 14.
20. G. Ginn, 'Answering the "Bitter Cry": Urban Description and Social Reform in the Late-Victorian East End,' *The London Journal* 31, no. 2 (2006).
21. C. Topalov, 'The City as Terra Incognita: Charles Booth's Poverty Survey and the People of London, 1886-1891,' *Planning Perspectives* 8 (1993), p. 411.
22. K. Bales, 'Charles Booth's Survey of Life and Labour of the People in London 1889–1903,' in *The Social Survey in Historical Perspective, 1880–1940*, ed. M. Bulmer et al. (Cambridge: Cambridge University Press, 1991), p. 68.
23. J.A. Riis, *How the Other Half Lives: Studies among the Tenements of New York* (New York: Charles Scribner's Sons, 1890), p. 25.
24. Dennis, *Cities in Modernity*, p. 98. The opera was first performed in 1947, adapted from a 1929 stage play by Elmer Rice, who based his descriptions on a tenement building on West 65th Street in New York.
25. D. Sibley, *Geographies of Exclusion: Society and Difference in the West* (London; New York: Routledge, 1995), p. 50.
26. Sibley, *Geographies of Exclusion*, p. 87. Bronislaw Geremek has similarly described how the city walls of medieval Paris were seen as a purifying device, defining territory within and without the walls, and placing the prostitutes beyond them. B. Geremek and J. Birrell, *The Margins of Society in Late Medieval Paris* (Cambridge: Cambridge University Press, 2006).
27. There are several competing explanations for this commonly accepted etymology. Benjamin Ravid points out that: 'while the word *ghetto* had never been applied to a Jewish quarter prior to 1516, compulsory, segregated and enclosed Jewish quarters had existed prior to 1516 in a few places' – some as early as the Frankfurt Jewish quarter, established in 1462. Benjamin C. Ravid, 'From Geographical Realia to Historiographical Symbol: The Odyssey of the Word Ghetto,' in *Essential Papers on Jewish Culture in Renaissance and Baroque Italy*, ed. D.B. Ruderman, Essential Papers on Jewish Studies (New York and London: New York University Press, 1992), p. 381.
28. R. Sennett, *Flesh and Stone – the Body and the City in Western Civilization*, first ed. (New York: W.W. Norton and Company, 1994), p. 215. While it is reasonable to assert that the Jews of Venice were foreign (as were many of its other inhabitants, due to the city being an important commercial port at the time), only a few decades after the formation of the Jewish Ghetto of Venice in 1516 the Jews of Rome suffered a similar fate. In this case they could not be defined as foreign in the same way as those in Venice; many of Rome's Jews had been there since the time of the Roman Empire.
29. In contrast, some records indicate the use of a secret channel to travel to the Jewish cemetery to avoid attacks on the funeral cortege.
30. Diary entry by John Evelyn from 15 January 1645, in *Memoirs of John Evelyn . . . Comprising His Diary, from 1641 to 1705–6, and a Selection of His Familiar Letters* (London: Henry Colburn), pp. 213–14.
31. An early source describes the Rome Ghetto at Easter: 'The Jews in this city are indulged in the use of synagogues; but are obliged to live all together in the Ghetto, as they call such places in the cities of Italy. At nine o'clock every evening the gates of the place where they live are shut up, and opened again in the morning; but at Easter they are locked up from Thursday in passion-week 'till the Monday following, during which time no Jew dares to be seen abroad. J Northall, *Travels through Italy: Containing New and Curious Observations on That Country* (London: Hooper, 1766), p. 128.
32. See Sennett, *Flesh and Stone*, p. 215.
33. D. Calabi, *Venice and Its Jews: 500 Years since the Founding of the Ghetto*, trans. Leonore Rosenberg (Milan: Officina Libraria, 2017), p. 63.
34. Ravid, 'Geographical Realia to Historiographical Symbol'.

35. D. Volchenkov and P. Blanchard, 'Ghetto of Venice: Access to the Target Node and the Random Target Access Time,' *Physics and Society* (2007), https://arxiv.org/abs/0710.3021.
36. S. Psarra, 'A Shapeless Hospital, a Floating Theatre and an Island with a Hill: Venice and Its Invisible Architecture,' in *8th International Space Syntax Symposium*, edited by M. Greene (Chair), J. Reyes and A. Castro, 016:011–016:028 (Santiago, Chile: Pontificia Universidad Católica de Chile (PUC), 2012).
37. It is notable that in the case of Rome the ghetto was located centrally, with an artery running through it, benefiting the commercial transactions between the ghetto's inhabitants and other, freer city inhabitants. Nevertheless, as Emily Michelson has pointed out, the Jews of the Roman ghetto were also endangered by the presence of Christians in their territory, due to accusations either of fraternisation or of inducements to conversion. E. Michelson, 'Exiting the Roman Ghetto,' in *The Ghetto: From Venice to Chicago*, ed. D. Feldman, B. Cheyette and F. de Vivo (Birkbeck, University of London: Pears Institute for the Study of Antisemitism, 2017). One explanation for this more central location may be the specific purpose of the Roman ghetto (established after Venice's, in 1555), which was to 'doubly fortify Catholic orthodoxy' by removing the Jewish population of the city (who in ancient times had enjoyed commercial freedom) from the public sphere and giving physical boundaries to the non-Catholic residents of the city. E. Michelson, 'Conversionary Preaching and the Jews in Early Modern Rome,' *Past & Present* 235, no. 1 (2017), p. 77.
38. Calabi, *Venice and its Jews*, pp. 32–3. She goes on to write: 'Daily life was enriched by shops for basic goods and workplaces. There was a bakery selling leavened and unleavened bread in Ghetto Nuovo and one in Ghetto Vecchio; there were numerous shops of fruit and vegetables, wine, meat, cheese, pasta, and wax candles. There was a barber, a hatter, a mender, a woman educating Jewish girls, a tailor, a bookseller (probably with an attached print and bookbinding shop); a workshop making products for alchemy. And there was an inn for foreign Jews as well as a carver, a lumber warehouse, another for tiles, and storage space for coffins.'
39. Sibley, *Geographies of Exclusion*, p. 57. For more on Baron Haussmann's reconfiguration of Paris, see section on the contemporary mapping of disease, in Chapter 2 of this volume (p. 49).
40. R. Powell, 'Loïc Wacquant's "Ghetto" and Ethnic Minority Segregation in the UK: The Neglected Case of Gypsy-Travellers,' *International Journal of Urban and Regional Research* (2012), pp. 120–1.
41. H. Mayhew, *London Labour and the London Poor* vol. 4 (London: Griffin, Bohn, 1861); Penguin Classics reprint edition, ed. V. Neuburg (Harmondsworth: Penguin, 1985), pp. 299–300.
42. R. Evans, 'Rookeries and Model Dwellings: English Housing Reform and the Moralities of Private Space,' in *Translations from Drawing to Building and Other Essays*, ed. R. Evans (London: AA Documents 2, 1997, first published 1978). 'Rookery' is a term dating from the 1820s to describe a particularly low quality of housing, closely packed with people of the poorest class (it refers to colonies of rooks, which nest in trees in large populations during the breeding season).
43. Evans, 'Rookeries and Model Dwellings'.
44. F. Engels, *The Condition of the Working Class in England* (New York and London: Panther Edition, 1891 (first published Leipzig: Otto Wigand, 1845), p. 57.
45. A. Mayne, *The Imagined Slum: Newspaper Representation in Three Cities 1870-1914* (Leicester: Leicester University Press, 1993), p. 161. Such language was commonplace at the time. The writer Henry James described the 'swarming' of Jewish immigrants in New York's Lower East Side in his travel essay on New York and the Hudson, where he notes the 'ubiquity of the alien'. H James, *The American Scene* (1907), p. 117–18.
46. Mayne, *The Imagined Slum*, p. 158.
47. For detailed analysis of the San Francisco map, see Chapter 5.
48. C.H. Nightingale, *Segregation: A Global History of Divided Cities* (Chicago: University of Chicago Press, 2012), p. 152.
49. See E. Durkheim, *The Division of Labour in Society (English Edition, 1964)*, trans. G Simpson, English ed. (New York: Macmillan Publishing Company, 1893); and Introduction to Hillier and Hanson, *The Social Logic of Space*, Cambridge: CUP, 1990. The argument set out in this passage was first published in a conference paper co-authored with Sam Griffiths: S. Griffiths and L. Vaughan, 'Mapping Spatial Cultures: The Contribution of Space Syntax to Research in Social and Economic Urban History,' paper presented at the Meeting of the European Association of Urban Historians, Helsinki, 24–27 August 2016.

50. As indeed was first pointed out by J. Hanson and B. Hillier, 'The Architecture of Community: Some New Proposals on the Social Consequences of Architectural and Planning Decisions,' *Architecture et Comportement/Architecture and Behaviour* 3, no. 3 (1987).
51. The fundamental space syntax theoretical argument for the agency of space is laid out in a series of papers by Bill Hillier: B. Hillier, 'In Defence of Space,' *RIBA Journal*, (1973); 'The Common Language of Space: A Way of Looking at the Social, Economic and Environmental Functioning of Cities on a Common Basis,' *Journal of Environmental Science* 11, no. 3 (1998); 'A Theory of the City as Object: Or, How Spatial Laws Mediate the Social Construction of Urban Space,' *Urban Design International* 3–4, no. 127 (2002).
52. This book does not require an understanding of space syntax to follow its argumentation. Nevertheless, the appendix contains an abbreviated explanation of its basic methods with a worked-through case using Charles Booth's 1889 maps of poverty as an example.
53. Hanson and Hillier, 'Architecture of Community,' p. 265.
54. See B. Hillier and L. Vaughan, 'The City as One Thing,' *Progress in Planning*: special issue on the syntax of segregation, ed. L. Vaughan, 67, no. 3 (2007), pp. 205–30.

2
Disease, health and housing

A double row of five-story tenements, back to back under a common roof, extending back from the street two hundred and thirty-four feet, with barred openings in the dividing wall, so that the tenants may see but cannot get at each other from the stairs, makes the 'court'. Alleys — one wider by a couple of feet than the other . . . skirt the barracks on either side. Such, briefly, is the tenement that has challenged public attention more than any other in the whole city and tested the power of sanitary law and rule for forty years . . . in the public records it holds an unenviable place. It was here the mortality rose during the last great cholera epidemic to the unprecedented rate of 195 in 1,000 inhabitants.[1]

The city as organism

Historical cartographers claim that the earliest use of maps to trace the spread of disease was in the United States in the 1790s, where they were used to record the location of infected households (see Figure 2.1), although for many centuries travellers have used maps to record the spread of exotic diseases amongst so-called native populations. In fact, some historians cite Aristotle's *De Aere, Aquis et Locis* ('On Airs, Waters, and Places', written *c.*400 BCE) as the first to assess climatic, topographical and geographical conditions to seek patterns of disease in local populations.[2]

Later examples include quarantine maps, such as that used to record the path of the plague of 1347–8. The resulting map was shown to correspond to trade routes, leading the government of Venice to halt all trade in its port for 40 days.[3] In later periods *cordons sanitaires* around diseased areas were determined by mapping the extent of a plague-ridden area in order to isolate it from areas that were disease free.[4] By the nineteenth century doctors were seeking to improve their practice by searching for environmental regularities in socio-medical phenomena by plotting them on maps. Such maps were used by doctors to look for causal relationships with environmental factors that may have caused disease. By plotting disease on the map, they were considering both spatial location and interrelationships, whether connections between data points or connections between the data and external factors, such as spatial location, social conditions, exposure to pathogens and so on.[5]

The metaphor of the city as organism can be dated back to the early nineteenth century, when the city was seen to be akin to the human body, wherein the circulation of water and air and the removal of waste created a healthy bodily state. Sanitary reformers took this further, wishing to use the mapping of disease as a method for identifying the causes of that disease. Taking a step beyond a medical topography, the mapping of disease sought to diagnose the city's disorder, a form of 'anatomical diagnosis of urban circulation and homeostasis'.[6] It is unlikely to be a coincidence that the earliest maps of urban social ills were those of disease, given that in the early days of medical geography the city itself was seen to be part of the problem of disease. Looking at one of the earliest examples (see Figure 2.1), in his inquiry into the cause of the prevalence of yellow fever in New York, the physician Valentine Seaman mapped the addresses of recent yellow fever fatalities in New York, 1797, observing 'that no Yellow Fever can spread, but by the influence of putrid effluvia', namely the accumulation of stagnant water and the close proximity of a sewer drain, which were viewed by Seaman as a likely source of the disease.[7] This sort of association of the disorder of the city with cases of disease was to continue for many decades further, until the first scientific diagnosis of contagion.

The use of maps to record disease was particularly common in early medical science, when, prior to Pasteur's work in the 1850s – which was to find the biological causes of disease – environmental causes were the focus for investigating disease.[8] One study has found that as many as 53 cholera maps had been published by 1832.[9] Yet the use of cartography remained limited to making visual correlations, instead of analysing the possible underlying factors, such as the physical (water quality or climate) or human (which would encompass man–environment relationships, as well as population size and composition).

In the United Kingdom, public concerns about outbreaks of the highly contagious disease of cholera in 1831, and subsequently 1848, 1854 and 1866, led to a series of reports which used maps to show the spread of

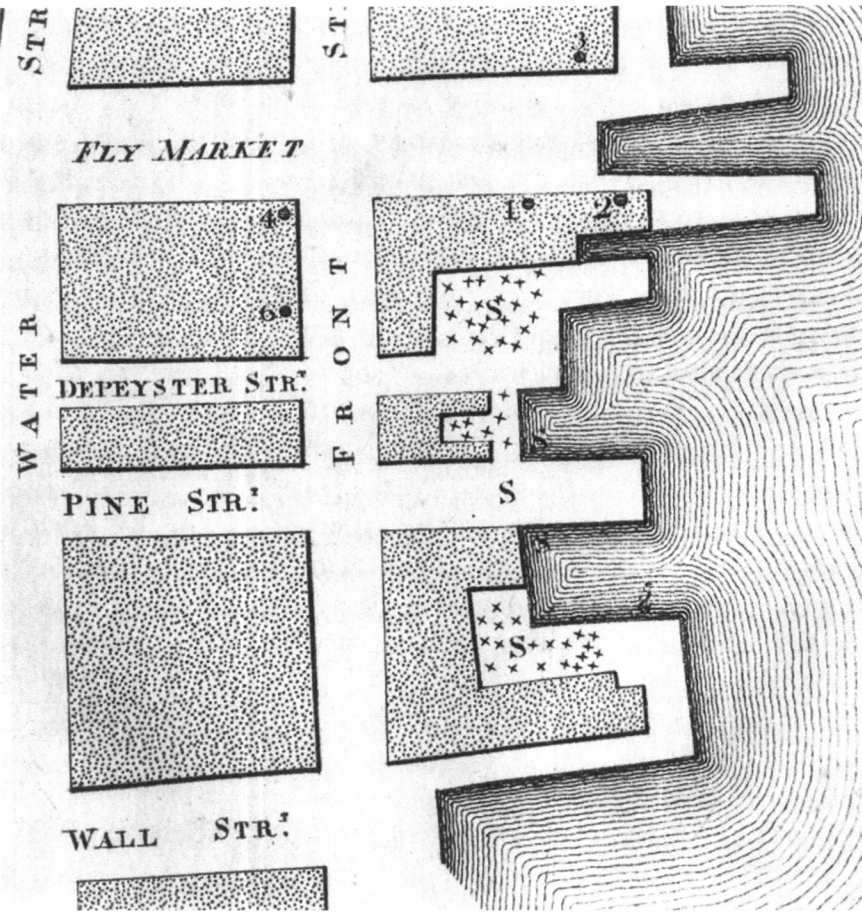

Figure 2.1 Detail of Plate II from Seaman's *An Inquiry into the Cause of the Prevalence of the Yellow Fever in New-York*, 10 March 1797.

Medical Repository, 1 (1800, 2nd edition): 303–323 [Rare Books Collection]. Image copyright Princeton University.

disease.[10] These included a report from 1849 by Shapter, which contained a lithograph map that used three different red symbols to identify deaths from the disease in 1832, 1833 and 1834.[11] The map also showed the city's response to the epidemic, such as locations where contaminated clothes were burned and buried, convalescent homes, druggists, burying grounds and soup kitchens. The notes on the map mention that Shapter believed cholera to be more prevalent in 'low-lying areas of dense habitation, near the river, where drainage was poor and waste and refuse accumulated – in other words, the disease was miasmatic' (namely, came from bad air).[12] A later map by Acland used black contour lines to mark altitude, to analyse the association between the prevalence of cholera and low-lying lands. Acland observed how the map of Oxford shows the infection to be 'concentrated in the undrained parts'. Somewhat anticipating the later work by John Snow in London, Acland also noted the significance of wells being contaminated 'by the sewage of the Town'.[13]

In the United States similar developments in medical cartography led to a study of cholera in 1850s Boston, where an outbreak of cholera led the city's leaders to commission a report that located all known hospital admissions from the disease in the city. The analysis also noted places of origin, age and meteorological reports against each case, while the map itself allowed its authors to take account of both the natural and human environment.[14]

We can see an early example of medical cartography in Edwin Chadwick's 457-page *Report on the Sanitary Conditions of the Labouring Population of Great Britain* from 1842. The report's maps of Leeds (see Figure 2.2) and Bethnal Green in East London are an interesting development in social cartography, in that they bring together three sorts of graphics: data on individual cases of disease (with cholera in red dots and respiratory disease in black) plotted on the maps; data on poverty levels, presented in three shades of block infill, with darkness increasing with poverty; and marking on the map (in a wash of brown colour) the areas containing streets lacking in 'cleansing'.[15] Notably, disease is presented here as being associated with dirt (and the choice of the colour brown is unlikely to be a coincidence). The graphic sophistication of these maps was insufficient to take the analysis much farther than associating dirt with disease, though the report to the Leeds Board of Health following the cholera epidemic of 1832 had attributed a lack of drainage, sewerage and paving as likely causes of the disease.

By the mid-nineteenth century, Leeds had become a centre for the manufacturing industry, with engineering and textiles the most dominant sectors. The city's natural resources, transport connections and regional location helped encourage rapid industrialisation.[16] In his map of cholera in Leeds, Chadwick identifies two groups of dwellings: houses of the working class and 'shops, workhouses, and houses of tradespeople'.

> By the inspection of a map of Leeds, which Mr. Baker has prepared at my request, to show the localities of epidemic diseases, it will

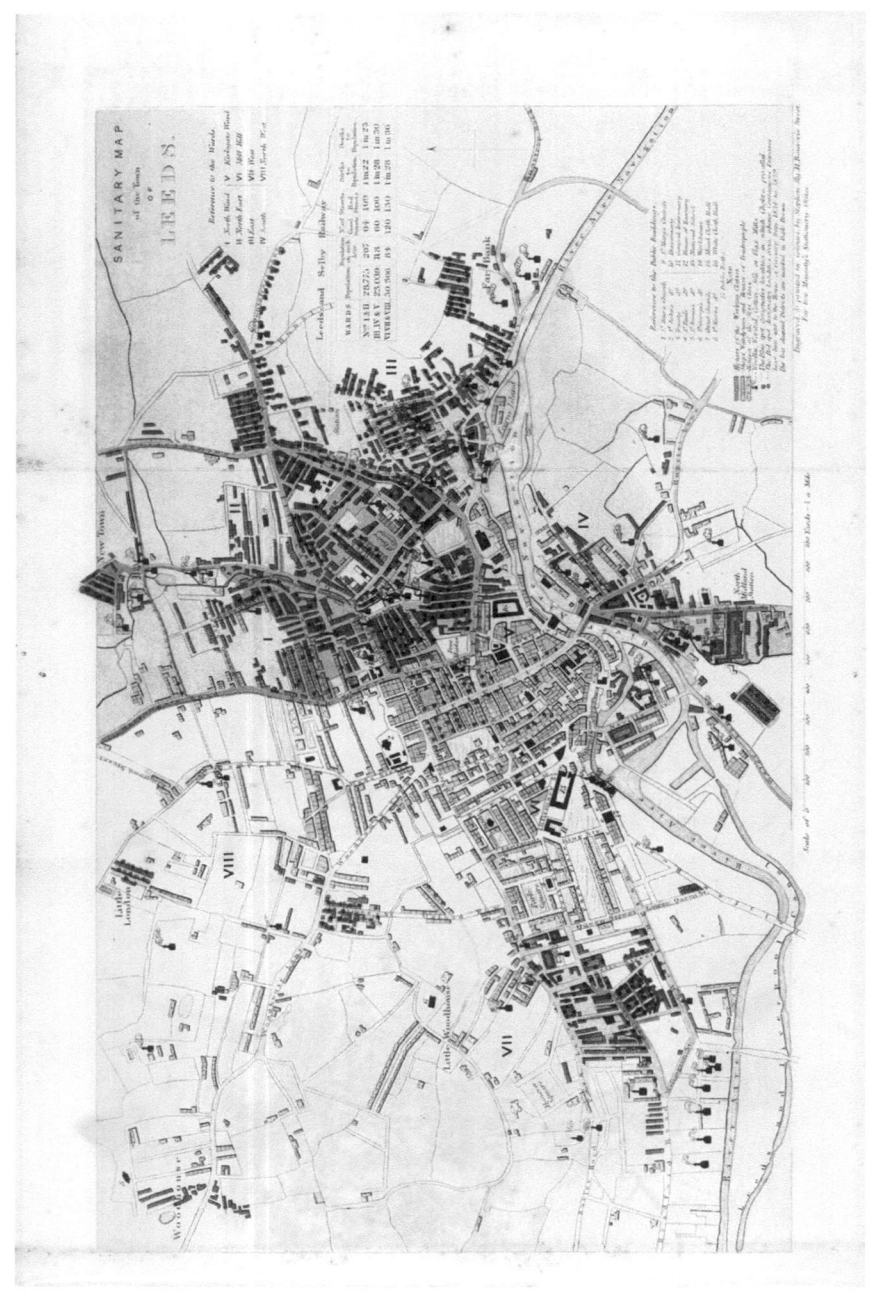

Figure 2.2 *Sanitary Map of the Town of Leeds*, 1842.
Copyright Cornell University – PJ Mode Collection of Persuasive Cartography.

be perceived that they similarly fall on the uncleansed and close [sic] streets and wards occupied by the labouring classes; and that the track of the cholera is nearly identical with the track of fever. It will also be observed that in the badly cleansed and badly drained wards to the right of the map, the proportional mortality is nearly double that which prevails in the better conditioned districts to the left.[17]

Dots (blue for cholera and orange for other contagious diseases) proliferate in the working-class areas. These are not contiguous regions but are sprinkled around the map. The map shows the importance of the river in shaping the land use pattern of the city, with residential areas dominating the area to its north and industries dominating the area south of the river. In fact, many of the city's industries utilised the river as a source of energy and transportation. The area north of the river and around the Leylands (the principal area of Jewish settlement in 1881) was primarily residential, but also contained smaller-scale industry such as leather works, shoe manufacturing workshops and small-scale brick factories.

Given the clear coincidence between disease and poverty, housing conditions were one of the most central concerns of the report, whose conclusions emphasised the need to improve the layout of working-class housing. While theories regarding the transmission of cholera through contaminated water were not then widely known, the relationship was becoming increasingly obvious to those who studied the map data closely:

> ... the public loss from the premature deaths of the heads of families is greater than can be represented by an enumeration of the pecuniary burdens consequent upon their sickness or death... The primary and most important measures [for improving health], and at the same time the most practicable, and within the recognized province of public administration, are drainage, the removal of all refuse of habitations, streets, and roads, and the improvement of the supplies of water.[18]

Space syntax analysis of the historical pattern of accessibility across Leeds's street network has found that the city's most accessible streets were situated north of the river and railway lines. Nevertheless, the sanitary map of Leeds (see section of the map in Figure 2.3), reflects the spatial conditions of the district, so although the Kirkgate area highlighted

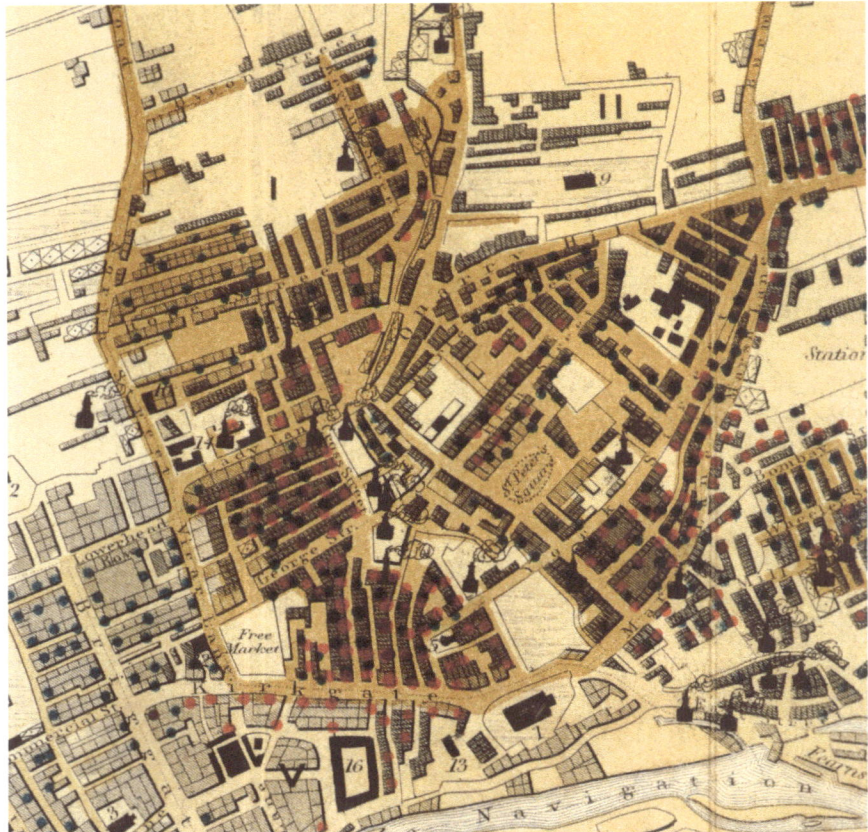

Figure 2.3 Detail showing the Kirkgate area of the *Sanitary Map of the Town of Leeds,* 1842. Copyright Cornell University – PJ Mode Collection of Persuasive Cartography.

here was quite close to the heart of the city, its northern part, known as Leylands, suffered from being inaccessible, with the main streets of the area skirting the district but generally not penetrating it.[19]

The records show that the Leylands was a very poor district with mostly back-to-back houses (see Figure 2.4) in cobbled streets and many tumbledown yards. A report from *The Lancet* from as late as 1888 found that many of the houses in the Leylands had no backyards or rear windows, and the housing density in the area was very high.[20] Aside from the cemetery, there was no open ground in the area and only in 1888 was the first recreation ground established in the district. In later years the area south of Roundhay Road had better housing, having benefitted from a bye-law of 1866 that required blocks of no more than eight

Figure 2.4 Back-to-backs on Nelson Street, Leeds, c.1901.
Leeds Library and Information Service, http://www.leodis.net.

terraces, with privies in between each block. Yet this housing was still of miserable quality and improved only in the 1890s, when several slum clearance programmes were implemented.

On the other hand, despite its accessibility, the southern area contained the worst housing in the district, ranging from back-to-back housing (with consequent seriously reduced ventilation and poor sanitation) to more orderly housing in the Regent Street and York Road areas, following the 'graph paper' terrace model.[21]

Elsewhere mapping of cholera was undertaken in Hamburg, Germany, by Dr Rothenburg, using the choropleth method to create hand-coloured gradations of red to display the relative aggregate incidence of the disease in 1832.[22] This map was reprinted in an 1850 British parliamentary report on cholera, helping to reinforce how widespread cholera was as a problem. Dr Hellis' map of cholera in Rouen, also in 1832 (see Figure 2.5 and Figure 2.6) was similarly influential. Published in his 1833 book, the map plotted each cholera case as a red dot on a highly detailed map, allowing for a visual inspection of the relationship between the location of shipping crew and clusters of disease.[23] Red lines were also used to

Figure 2.5 *Cholera in Rouen,* 1832.
Eugène-Clément Hellis. Reproduction courtesy of Emmanuel Eliot and the Geoconfluences project.

outline establishments such as the prison or army barracks. The report also records the ports of origin of the ships. His multiple observations led Hellis to the conclusion that cholera dispersed along the banks of the river and into particularly impoverished and run-down dwellings, so starting, at least in principle, to favour a contagion theory.

A recent study assessed the Hellis map by modelling the two main factors assumed to be responsible for propagating cholera: the presence of an aquatic environment (river and wells) and the density and level of income (measured by charitable expenditure).[24] The computer simulation found a close association between health risk according to these spatial and social determinants and the clusters of high rates of cholera. This study shows that with sophisticated analysis it is possible to consider socio-spatial relationships in great detail. In this instance, the authors provide evidence that places of exchange (such as markets), the quality of the housing, poverty and – importantly – water supply, were all likely to

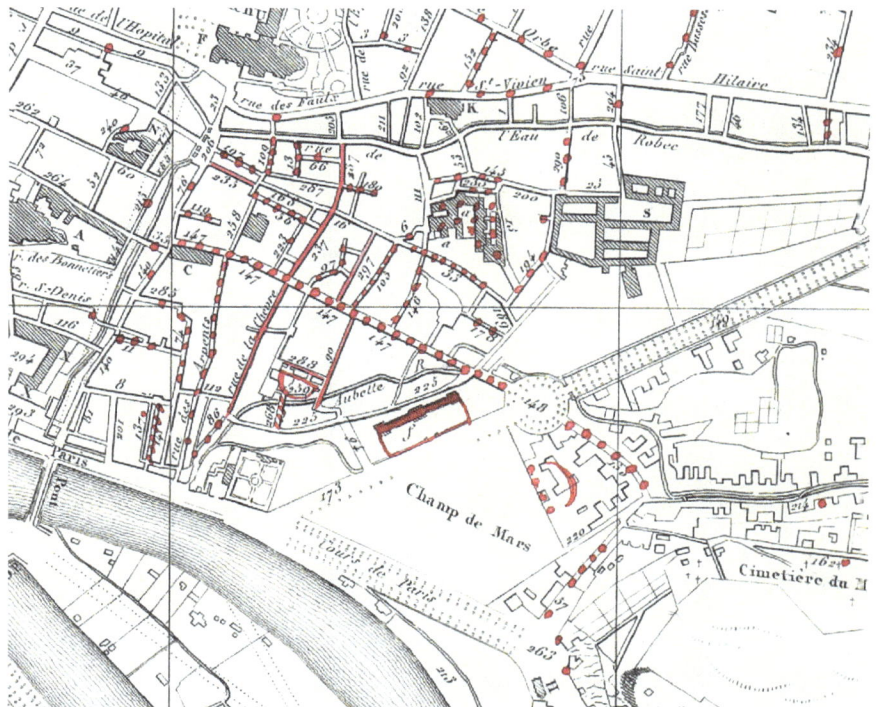

Figure 2.6 Detail of *Cholera in Rouen,* 1832.
Eugène-Clément Hellis. Reproduction courtesy of Emmanuel Eliot and the Geoconfluences project.

have contributed to the spread of the disease over and above the fact that its original source may have been seafarers coming into the city.

By the time of the cholera outbreaks in 1830s and 1840s London, the streets of its district of Soho had become a crowded slum that had gone down in the world. It was a place that Dickens' Nicholas Nickleby found had

> ... pieces of unreclaimed land, with the withered vegetation of the original brick-field. No man thinks of walking in this desolate place, or of turning it to any account. A few hampers, half-a-dozen broken bottles, and such-like rubbish, may be thrown there, when the tenant first moves in, but nothing more; and there they remain until he goes away again: the damp straw taking just as long to moulder as it thinks proper: and mingling with the scanty box, and stunted everbrowns, and broken flower-pots, that are scattered mournfully about – a prey to "blacks" and dirt.[25]

In fact, space syntax analysis of the built form shape and configuration of the district of Soho in the latter part of the nineteenth century has found that the area was significantly more spatially segregated then its surrounding streets. This can be seen in Figure 2.7, which shows a section of the space syntax analysis for the Soho area c.1890 overlaid on the Ordnance Survey map of the area from the same period. Oxford Street, running south-west to north-west, appears in the warmest colour, signifying its wide-scale accessibility, while the area south of that street is markedly more segregated (note cooler shades of blue and turquoise surrounding Soho Square, south-east on the map).

Further analysis found that a combination of shorter blocks and a complex morphology, with narrow visual fields from the main streets surrounding the area into its interstices, meant that flows of movement into the area would have been restricted. The result was that the Soho district became socially separated from its prosperous surroundings,

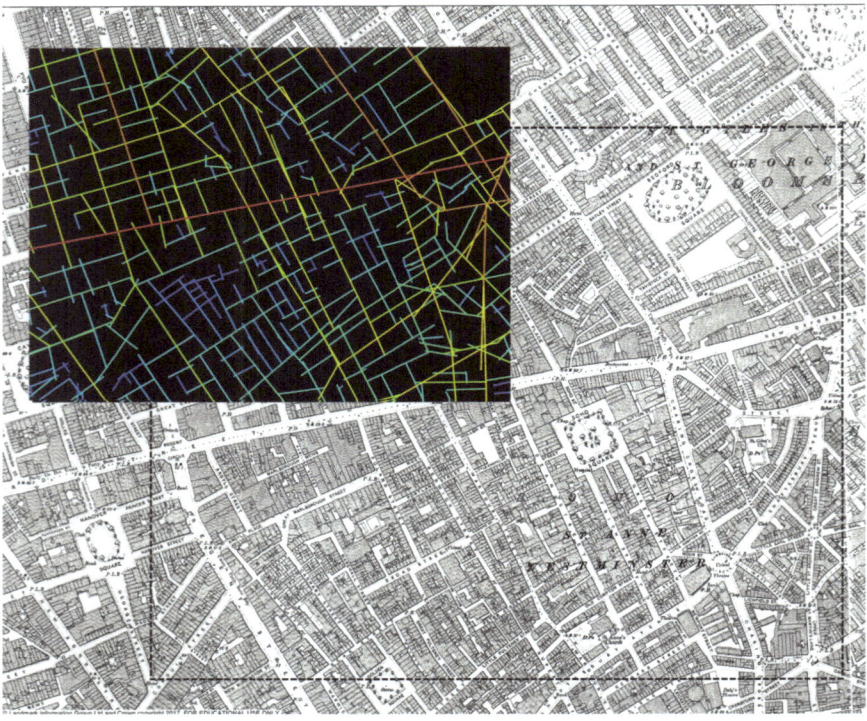

Figure 2.7 Space syntax analysis of the Soho area c.1890 showing local accessibility (axial radius 3) (inset) and Ordnance Survey map of London of the same period.
Image by the author.

attracting itinerant migrant workers, such as the seasonal Italian ice-cream sellers of the nineteenth century, radical revolutionaries at the turn of the twentieth century, or fashion sub-cultures in the twentieth and twenty-first centuries.[26]

At the same time, Dr Perry, Senior Physician to the Glasgow Royal Infirmary, was one of many physicians contributing to the plethora of medical reports on disease and housing conditions. In a study into the extent of the typhus epidemic in Glasgow of 1843 he pinpointed individual households affected by the fever. Using reports collected from the local surgeons he had a team colour up a map of the city, shaded darker where the epidemic was most prevalent. His analysis showed six-fold differences between parts of Glasgow when calculating fever cases as a proportion of the population, with a clear association between overcrowding and disease – but also vice.

> . . . those places most densely inhabited, by the poorest of the people, have suffered most severely. The epidemic, having once got into a densely crowded land or close, never ceased until it had visited every house, and in many of the houses every inmate . . . The houses in most cases were too crowded; some small apartments containing two, three, and occasionally four families! In others numerous lodgers, men, women, youths, and children, were huddled indiscriminately together on a cargo of straw upon an earthen floor . . . a fruitful source of vice, pollution, and disease.[27]

Medical mapping as statistical method

It was not until John Snow published the second edition of his essay 'On the Mode and Communication of cholera', showing the association between the incidence of deaths from cholera and the location of pumps in the Soho district of London, that mapped statistics were used scientifically (see Figures 2.8 and 2.9).[28] London was among many large cities that had been suffering from periodic cholera epidemics since the 1830s. Snow hypothesised that cholera was caused by a germ spread through contaminated water, in contrast with the prevalent miasma theory of contagion through bad air. The two theories were debated vigorously, but Snow's ground-breaking studies of the 1853–5 epidemic resulted in his being considered the father of epidemiology, due not only to his well-reasoned statistics but to his maps, which illustrated to

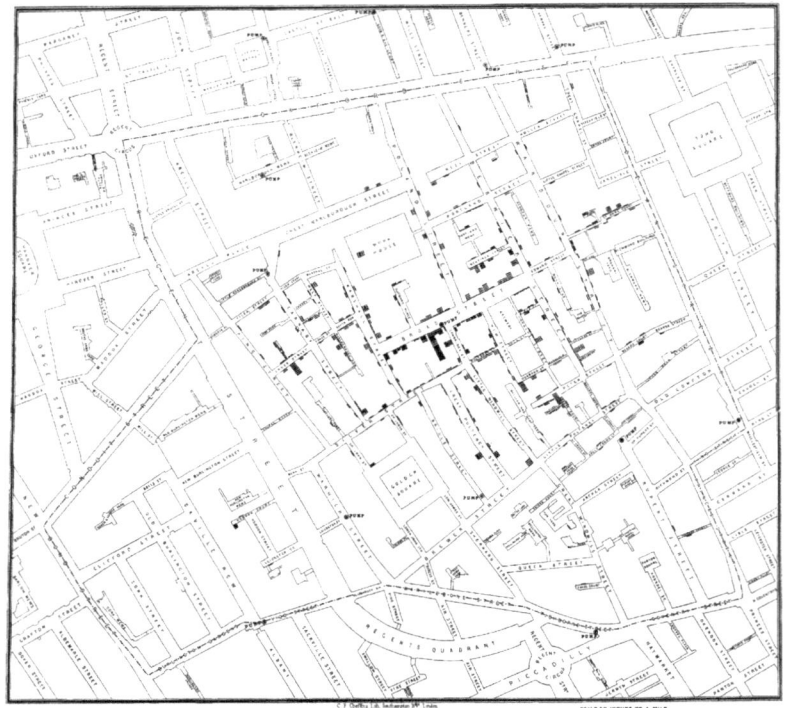

Figure 2.8 J. Snow, *Street Map of Soho, around Golden Square, Illustrating Incidences of Cholera Deaths during the Period of the Cholera Epidemic*, 1853.

Published in J. Snow, *'On the Mode of Communication of Cholera'*, 2nd ed. (London: John Churchill, 1855). Map courtesy Ralph R. Frerichs, UCLA Department of Epidemiology, School of Public Health.

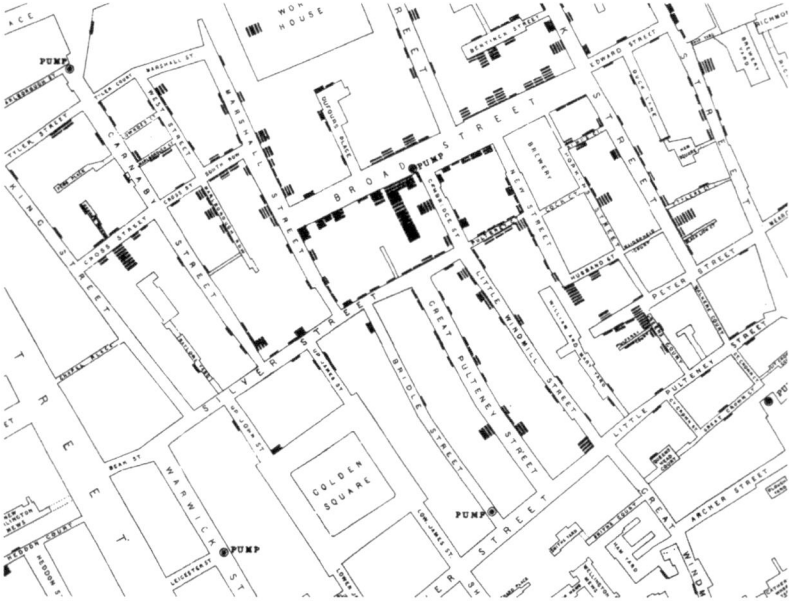

Figure 2.9 Detail of Fig. 2.8, J. Snow, *Street Map of Soho*, 1853.

a lay audience the evidence of a cluster of cholera cases amongst people living close to a single water pump on Broad Street and, hence, that contaminated water was the source of the disease (see Figures 2.8 and 2.9 showing John Snow's Broad Street map).

Notably, instead of using his map as a source of evidence, Snow used it to communicate that evidence. While some critique the mythology that surrounds the Snow story (which states that, following his testimony, the handle of the pump believed to be the source of the disease was removed), Snow's central role in the history of social mapping is significant. His importance lies first in his establishment of a clear spatial relationship between contaminated water and the disease, and second in his use of disease mapping to observe, communicate and analyse statistics. By using statistics to disprove the air theory, he pinned down the causation in such a way that his argument became incontrovertible. In order to understand the results of his study in their full complexity the reader had to take in a mass of statistical tables, which, once read alongside the maps, became clear; at the same time the clarity of the maps made them devices for communicating a simpler form of the argument to the general public.

Snow allowed the viewer of the map to – as Gilbert puts it – '. . . conceptualise [the causal relationships] in a whole new way, having to do with the actions of humans in relation to the environment'.[29] While the stories around Snow project a picture of a single pump affecting a single group of people, in fact his analysis demonstrated that people's pattern of illness was (unsurprisingly) much more complex, with some of the deaths close to the pump being due to contagion having taken place elsewhere; other deaths were recorded outside of the area, even though the people had been infected by contaminated water taken from the pump. Gilbert also points out that Snow's second map, which shows the coverage of the various water companies supplying the wider London area, emphasises this reading of a community as having a much wider spatial scale than the Broad Street map would suggest. While Dorling and Shaw argue that Snow's dot-mapping method is a problematic precedent, in that it is easily misread if it is not normalised for the size of the population; indeed, by centring the map on Broad Street, it emphasises the individual pump's culpability, when in fact there is evidence that while there were higher rates of cholera elsewhere in London at the time, cases were lessening overall due to the work of the local boards of health to clean up water supplies following the passage of the 1848 Public Health Act.[30]

Although he is less well known today, the importance of the work of William Farr in compiling the statistics on cholera mortality used by Snow has been highlighted by Tessa Cicak and Nicola Tynan. In their study, they used a Geographic Information System to map London's water company boundaries at that time. They then tested Snow's data: first by correlating them with figures on ground elevation above the level of the River Thames and then by correlating them with the water company supply areas. Their results substantiate Farr's proof of the 'influence of water as a medium for the diffusion of the disease in its fatal forms'.[31] Farr had initially believed that elevation was the strongest correlate with cholera mortality, but by 1853–4 he was persuaded by the statistically significant differences in mortality rates between different water companies, after which he switched allegiance to Snow's theory.

Snow's work remained influential for a considerable time onwards. While the waterborne theory of cholera became more commonly accepted, his map was viewed as a way to explain the impact of 'microorganisms invisible to the human eye', as Steven Johnson has put it.[32] Snow's longer-term impact was two-fold: in starting a tradition of medical mapping that continues to this day, but also in pushing for more accurate, large-scale maps of cities. His study also helped establish a standard cure for the spread of the disease, namely direct intervention in the environment, whether through sanitary improvements or wholesale redevelopment of the slums. These environmental solutions are part of a step-change that took place in this period, whereby medical and public health practitioners were able to construct a 'fine-grained, spatially precise argument', based on street-by-street data, namely data that corresponded to the environment within which the victims of the disease were living at the time.[33] Snow's map also serves as an exemplary research instrument, to 'illustrate an inferential process of deduction [to be employed] in considering the logical relationship between spatially grounded data on disease incidence and environmental [in this case] principally water-borne, sites of suspected contagion'.[34] In effect this practice continues to this day in epidemiology, which regularly uses maps to analyse spatial correlates of disease, of which more in the last section in this chapter.

On housing and health

By the mid-nineteenth century, massive growth in industrial towns, in the absence of stringent regulation of house building, road laying, water supply, and sewage removal, had led to increasing public concerns in Britain. The quality of housing, unmade roads and overcrowding became the focus of debate. Edwin Chadwick's 1838 report on living conditions in east London, as well as the report on the *Sanitary Conditions of the Labouring Population of Great Britain* in 1842, led to recommendations to build drains, remove sewage via sealed pipes, improve water supplies and build cemeteries on the edges of towns. Yet it was only with the passage of the 1848 Public Health Act (due partly to proof from William Farr that linked disease with squalor and to Chadwick's argument about the cost of death and illness for the nation's productivity) that improvements to the quality of housing and urban layouts started to take shape. Following Snow, the focus of public health efforts led to the appointment of John Simon as first Medical Officer of Health from 1858, who pushed prevention as well as curative methods. The Public Health Act of 1875 subsequently required local government bodies to improve water sanitation and to appoint inspectors responsible for housing standards, quality of waterways and so on.

As Pamela Gilbert has shown, the cholera epidemics that plagued London in the nineteenth century were a turning point in the science of epidemiology and public health, and the use of maps to pinpoint the source of the disease initiated an explosion in medical and social mapping not only in London but throughout the British Empire as well. She writes that 'medical mapping . . . is essentially a statistical argument presented visually . . . it also comes into being because of the spatialized understanding of social problems'.[35] Gilbert goes on to explore an anatomical metaphor for the way in which Booth coloured his maps, demonstrating a form of visual rhetoric in the way in which the map is coloured:

> . . . the west appears to be a healthy, bright orange-red with pockets of darkness. The east . . . and to some extent the south, range from a pallidly under-oxygenated pinkish-lavender to a chilling pale blue, with threateningly concentrated blotches of dark blue and black. Given the ethnic distribution of the London population, this darkness may have taken on racial overtones as well.[36]

Claiming that Booth's understanding of poverty was 'informed by nineteenth-century sanitary and medical understanding of both public and individual health', Gilbert sees Booth's dislike of dark, airless spaces as coming from a medical understanding of sanitation: '. . . we see again the notion of circulation which is encoded in the anatomical representation of London's body'.[37] Interestingly, this could be said equally to apply to the way in which Booth himself viewed the city, whose severing by railways could cut off the life-blood of its inhabitants: 'In Battersea poverty is caught and held in successive railway loops south of the Battersea Park Road... This is one of the best object-lessons in "poverty-traps" in London.'[38] See the reference to gangrene by Harold Dyos:

> A... careful reading of Booth's maps would show how some additions to the street plan – a dock, say, or a canal, a railway line or a new street – frequently reinforced these tendencies. What often made them more emphatic still was the incense of some foul factory, a gas-works, the debris of a street market, or an open sewer. They all acted like tourniquets applied too long, and below them a gangrene almost invariably set in. The actual age of houses seldom had much to do with it and it was sometimes possible to run through the complete declension from meadow to slum in a single generation, or even less.[39]

After the publication of Booth's maps there was increasing public pressure to clear out the poor areas and tidy up the existing intricate and enmeshed networks of streets. As we saw in the previous chapter, the city's morphology – its physical form and layout – was itself viewed by the general public as a source of the immoral behaviour of its inhabitants, and a significant obstacle to policing. We will see more on the importance of Booth's study in influencing housing and urban legislation in Chapter 3.

In the United States the housing tenements of densely occupied cities such as New York attracted a series of housing commissions and enquiries whose impact was limited until the turn of the twentieth century. Population growth in New York in the latter half of the nineteenth century had led to a rapid increase in population densities – a particularly acute problem on Manhattan island, especially within the constraints of the standard lot size of 25 by 100 feet. With this increase in population, the tenement building emerged as a response to the demand for low-cost housing design. Yet, without building regulation, buildings were constructed with a severe lack of light and ventilation. Up to 90 per cent of a lot was permitted to be built upon, producing a sequential layout akin to a railroad carriage (hence their being called 'railroad flats'). The 1867 and 1879 Tenement Acts sought to introduce legislation to improve housing and led to a new housing form, the 'dumb-bell' (so called because of its shape in plan – see Figure 2.10), which was an improvement, but did not yet achieve the desired outcome of improvements in light and ventilation. It still permitted up to 65 per cent coverage of each lot.

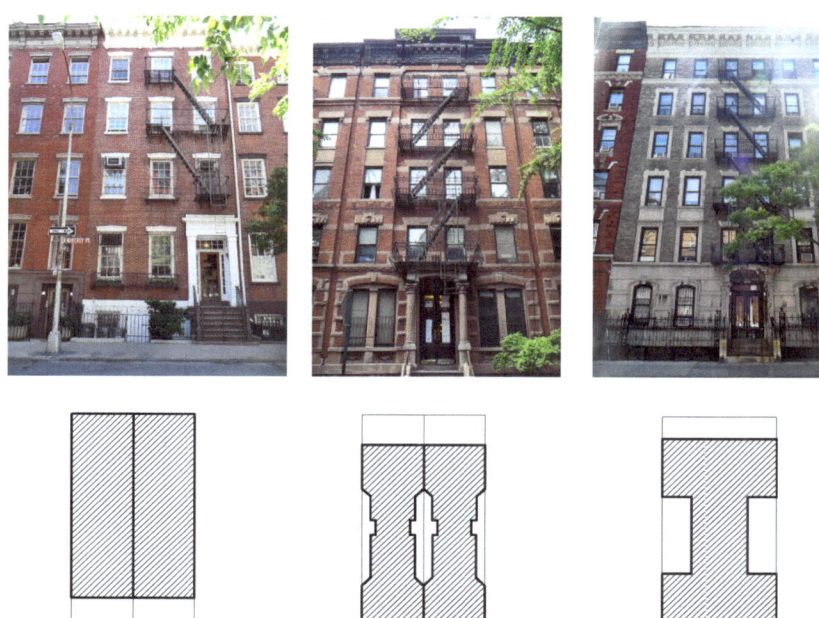

Figure 2.10 Photographs of tenement housing styles in New York: (from left to right) 'old-law' railroad-type housing, 'old-law' dumb-bell-type housing, and 'new-law' tenement housing as it appeared c. 2013, and the associated ground-plans.

Image courtesy Garyfalia Palaiologou, 2013.[40]

The Tenement House Committee was created by the New York State legislature to enquire into New York's housing problem. Its report, and maps, were presented on 17 January 1895 and sparked much public interest after they were published in *Harper's Weekly*, a widely read publication that had frequently written on poverty and housing problems. The maps 'represented an important milestone in the use of new forms of graphic representation by reformers'.[41]

Of the two maps, the upper shows population density, with the highest densities located in the Lower East Side, an area of high immigrant settlement (see photograph of New York tenement in Figure 2.11 and maps in Figure 2.12). The lower map was adapted from a colour version that was shown to the committee, and that, interestingly, anticipated Jacob Riis' conception of the city's various nationalities, which (as we saw in the Chapter 1),

> would show more stripes than on the skin of a zebra, and more colours than any rainbow. The city on such a map would fall into two

Figure 2.11 'Lodgers in a Crowded Bayard Street Tenement – "Five Cents a Spot"', Lower East Side, New York, 1889.

Jacob Riis, 1889, via Wikimedia Commons.

great halves, green for Irish prevailing in the West Side tenements and blue for Germans on the East Side. But intermingled with these grand colours would be an odd variety of tints that would give the whole the appearance of an extraordinary crazy quilt.[42]

Although they were never interpreted in detail, the maps were a powerful representation of two coinciding urban characteristics: on the one hand, the cluster of extremely high population density in one corner of lower Manhattan and on the other, this being the heart of the highly diverse immigrant quarter.

The Tenement House Committee's maps appeared almost simultaneously with several reports, which collectively helped speed along the programme to cut out the diseased areas of the city from its body. In one example, a ground-plan of a 'Lung-Block', a single block in New York that was riddled with cases of tuberculosis, was published in a report by Dr John Bessner Huber (Figure 2.13).[44] His report is full of graphic descriptions of the conditions in this single block of dwellings, whose

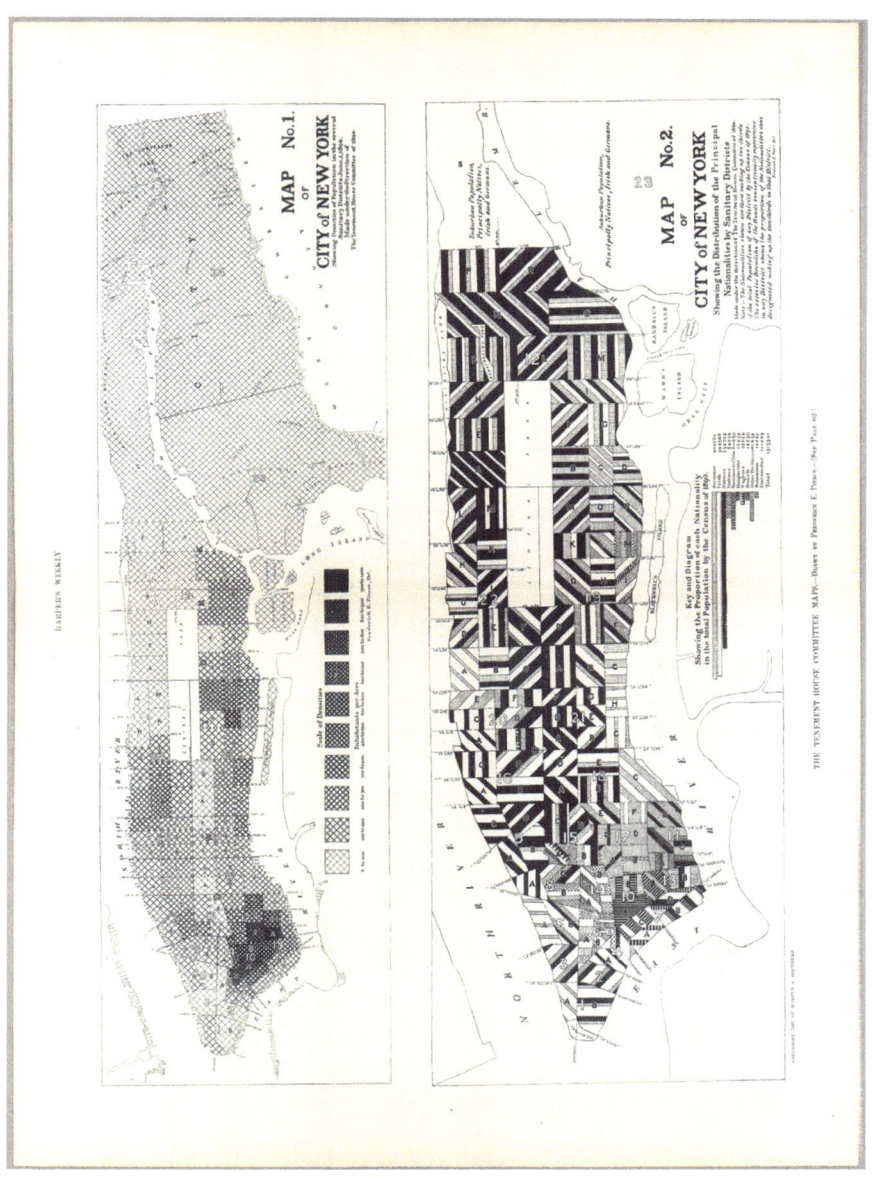

Figure 2.12 The Tenement House Committee maps, 1894.
Copyright Cornell University – PJ Mode Collection of Persuasive Cartography.[43]

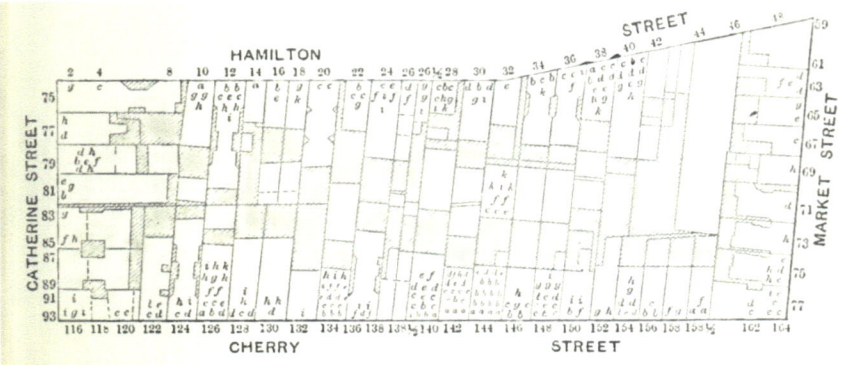

FIG. 37.—Ground-plan of the "Lung Block." The shaded sections are courts and air-shafts. Each letter represents one case of consumption reported to the Health Department since 1894. a = one case in 1894. b = one case in 1895. c = one case in 1896, and so on to k = one case in 1903. (As it is not possible from the records to tell whether a given case occurred in the front or rear tenement, all have been assembled in the front building, except in 144 Cherry Street, where there was not room.) In the plans of the Health Department (Part XII, Chapter 11) dots (.) take the place of letters.

Figure 2.13 The Lung Block, New York, 1903.

From J.B. Huber, *Consumption, Its Relation to Man and His Civilization, Its Prevention and Cure* (Philadelphia: Lippincott, 1906).[45]

appalling physical conditions he blamed for the block's extremely high rates of consumption (tuberculosis), such that,

> Infection comes not only from the room, but as well from halls and stairways. An old Italian, a hopeless victim, sits out on the steps in front all day long in the sun, while the children play around him, and all through the evening, with men and women beside him. His cough never stops. The halls behind and above are grimy, offensive, lying heavy with cobwebs, and these cobwebs are always black. The stairways in the rear house are low and narrow, uneven, and thick . . .[46]

Alongside various disease maps of this time, about which further discussion is presented below, the impact of this map led to reforms in housing and ultimately to the Tenement House Act of 1901, which followed many of the prescriptions of the Committee's report.

The graphic impact of Huber's ground-plan was widespread. It featured in *The Brooklyn Daily Eagle* newspaper in 1903, along with a long article discussing the appalling situation in blocks such as this.[47] The powerful image of the ground-plan, with its patches of shading to indicate the miniscule amount of open space in its interior, marked with the many cases of disease, crossed the continent: just a couple of weeks later the *Los Angeles Herald* described the building as 'supplied with tuberculosis germs on its walls and its ceilings, in its hallways, on all the furniture and in the dirt-filled cracks in the floors'.[48] The article goes on to blame the city for the fact that such buildings are permitted to stand, but notes also that a new ordinance – evidently the Tenement Act of 1901 – will hopefully eliminate such disease-breeding buildings. Indeed, soon after this date housing reform took shape in the form of localised slum clearance, which then became more widespread with a systematic programme of slum clearance and public housing construction. Only a few years later another doctor, this time in Chicago, made a comprehensive study of the incidence of tuberculosis in the Near West Side of Chicago. His report's striking graphics, which show building morphology and land use alongside mortality cases, were another step forward in using maps to test hypotheses regarding the causes of contagious disease (Figure 2.14 and detail in Figure 2.15).

The spatial solution for disease in the 'body' of the city shifted over time. By the end of the nineteenth century, Charles Booth was advocating suburbanisation as the best solution to 'the evils of over-crowding', proposing a system of tramlines to provide easy commuting routes that would allow London to be broken up into suburban centres. These went on to be constructed alongside a programme of widening thoroughfares, and courts were opened up to allow for a battle to be fought against 'the war with dirt, disease, and premature death.[49] At the same time in the United States, following the New York State Commission, other states picked up the issue of overcrowding, not only at building scale, but also at the scale of the lot or the block (in something of a recollection of the early Housing Acts of the city). In one example, a map of a Blind Alley in Washington and its associated report was explicitly attributing the lack of through passage as one of the causes of disease and crime in the city (see Figure 2.16).[50]

The Blind Alley map was published in a 'Directory of Inhabited Alleys in Washington' from 1912, which was drawn up to allow for easy inspection of these alleys.[51] The directory cites the death rate in alleys as exceeding that in streets by a considerable degree, with the most prevalent diseases being pneumonia and tuberculosis. The solution is also outlined in the directory; it cites the relevant District Codes which will allow the alleys and minor streets to be extended, widened or straightened for purposes of improving health. The poor health of the alley's inhabitants was attributed by the directory's authors to their living conditions – yet the

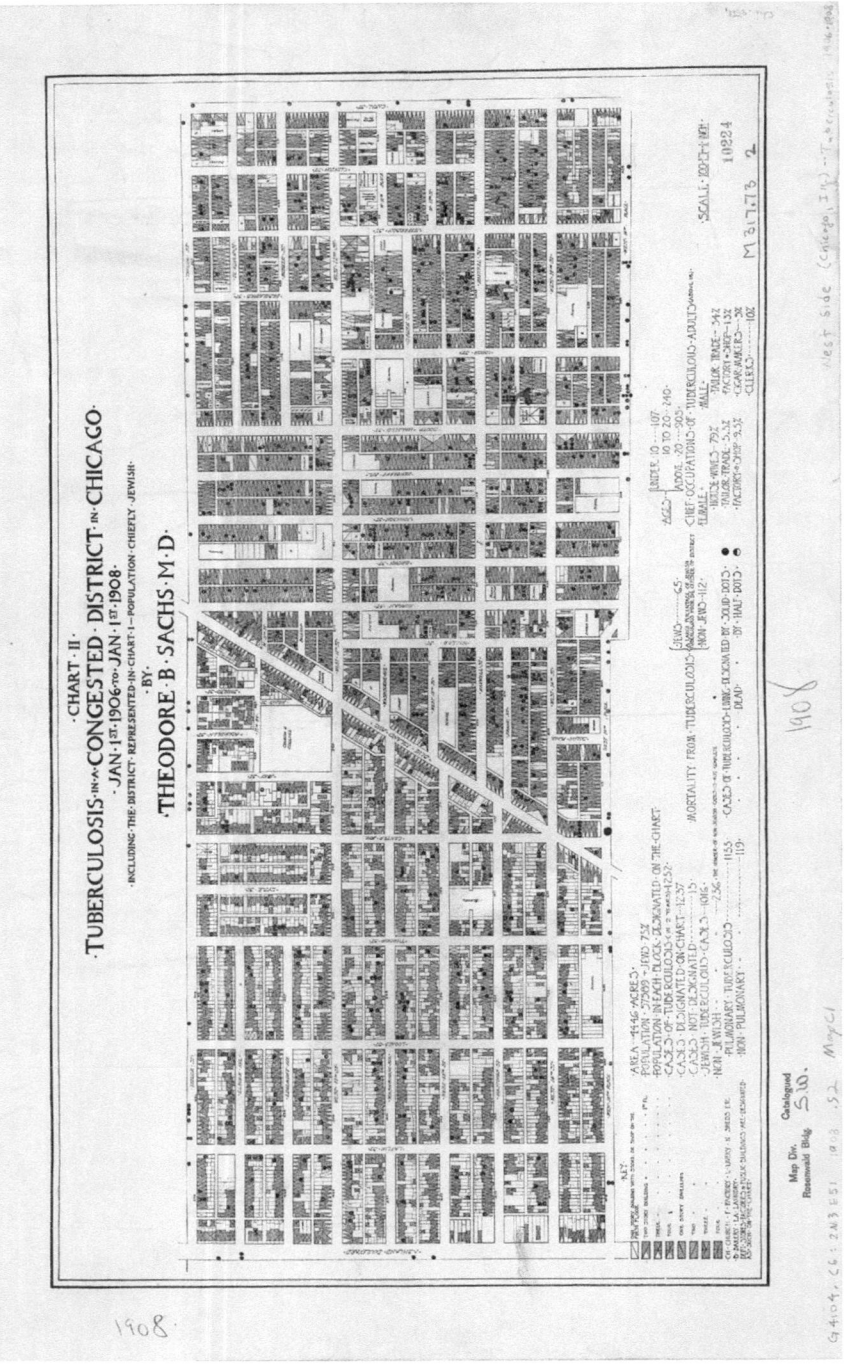

Figure 2.14 *Tuberculosis in a Congested District in Chicago, Jan. 1st, 1906, to Jan. 1st, 1908, including the district represented in chart 1, population chiefly Jewish.*

Theodore B. Sachs. Image credit: University of Chicago Map Collection.

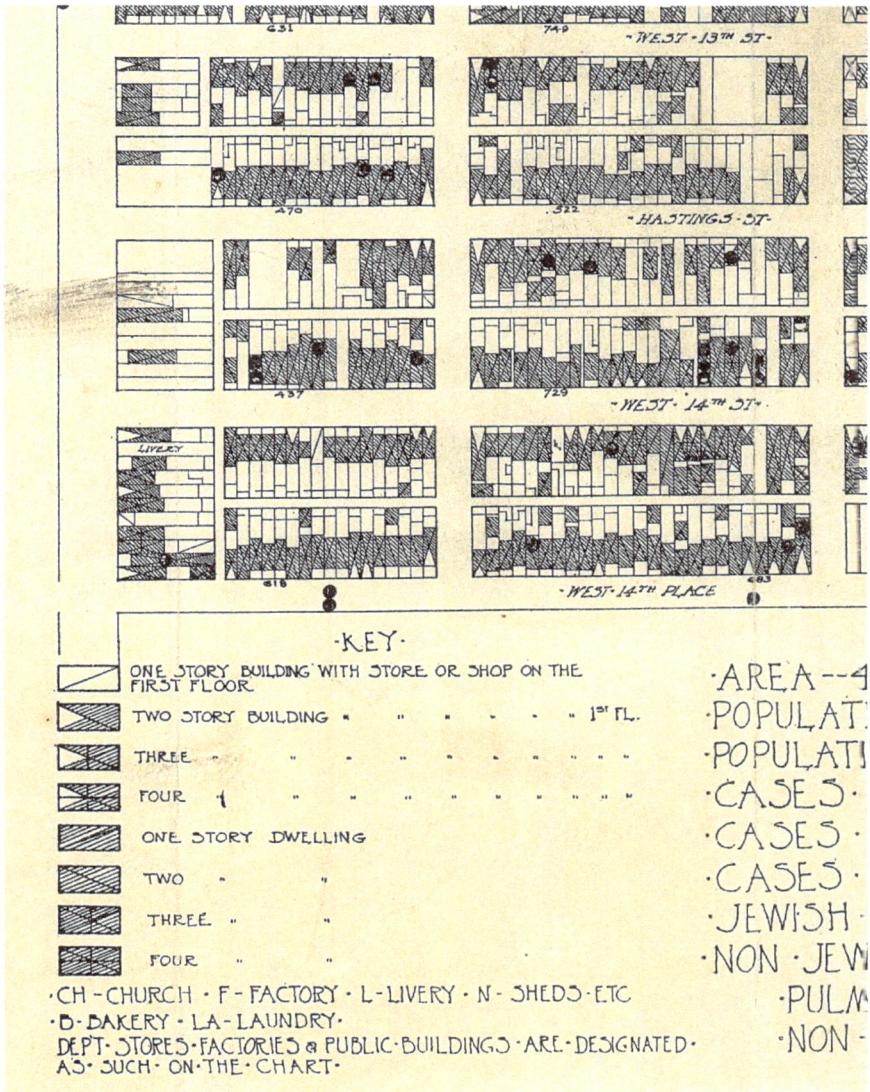

Figure 2.15 Detail of *Tuberculosis in a Congested District in Chicago*.
Theodore B. Sachs. Image credit: University of Chicago Map Collection.

analysis of *Alley Life in Washington* by James Borchert refutes this argument, describing how the Black American migrants from the Southern states had made the most of the layout of street layouts such as these to reinforce internal communal ties, creating a reciprocal relationship of support.[52]

DISEASE, HEALTH AND HOUSING   47

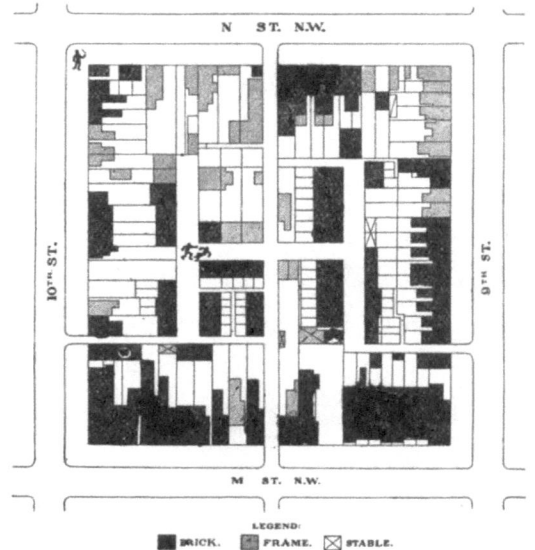

Figure 2.16 *The Blind Alley of Washington: Seclusion Breeding Crime and Disease*, 1911.
Image courtesy Jenny Schrader.

By the end of the nineteenth century major advances in bacteriology meant that a concern with the biological causes of illness – a long list of which includes tuberculosis, tetanus, dysentery and the old enemy cholera – shifted the focus from mapping disease to the new urban problem of the era: mass immigration. Not only was this seen to be intensifying poverty in urban areas, it was also coupled in the public mind with contagious disease, a concern reaching its peak when outbreaks of yellow fever became associated with Chinese migration to major cities in Australia, the US and the UK. One striking example of how fear of contagion led to a racialised mapping of San Francisco's Chinese quarter will be looked at in Chapter 5.

Following a decline in the use of disease maps in the latter part of the nineteenth century (to be supplanted by maps of ethnicity and poverty), a resurgence came in the twentieth century, when social planners and so-called 'hygienists' started to conduct surveys and then map cultural traits, such as language, dialect and custom. This chapter's last section demonstrates that disease maps are increasingly being used in contemporary public health research as tools both for diagnosing disease and for testing out cures for it. A wide variety of medical disciplines use disease mapping nowadays to analyse disease incidence and spatial distribution, frequently working in a multidisciplinary (and sometimes interdisciplinary) team that could include forms of expertise as varied as anthropology, transport and geography, although for the medical sciences the main practitioners come from epidemiology and public health.

Contemporary mapping of disease: a story of spatial inequality

While contemporary use of medical cartography continues the historical tradition of mapping disease spatially, medical mapping has shifted its focus in modern times to look mostly at non-contagious disease. Health cartography typically captures possible environmental factors, for etiological research (the study of the causes or origins of disease), as well as more complicated mapping of health (or wellbeing) and disease in relation to other factors, such as studies of social equity.

A classic example of etiological research is a study of the prevalence of Burkitt's lymphoma in the 1960s. Disease maps showed a wide variation in the number of cases, which seemed to correspond with climate: incidence was high in the damp, hot climatic conditions of the lowland areas of Uganda and the coastal regions of Kenya and Tanzania, and low in the highlands of south and west Uganda, central Kenya and northern Tanzania. Analysis of the maps and further research led to the conclusion that Burkitt's tumour is related to the presence of malaria (which is more prevalent in moist, hot regions).[53]

A western-centred review should also mention the continuing problem with contagious disease around the world, with cholera rearing its ugly head again in war-ridden Yemen at the time of writing this book. Indeed, just a few years ago, it took careful spatio-temporal analysis by a group of physicians to unpick the causes of an outbreak of the disease in Haiti, which proved to have been brought by United Nations peacekeeping troops from Nepal who came to the rescue of earthquake victims on the island.[54] The lack of access to clean water and decent sanitation meant that once the disease had been introduced, it spread rapidly.

Dr Snow's dot maps continue to be used to identify clusters of diseases such as cancer, as a way in which to identify possible local environmental causes. It is essential, however, to go beyond simply identifying problems by eye and embrace more sophisticated mapping and mapping analysis to avoid misattribution of cause and effect, as well as to avoid missing other causes that are not easily identifiable at first glance.[55]

In the context of this book, it is interesting to cast the gaze back to the so-called developed world,[56] to see how contemporary mapping methods allow us to understand the much more complex interplay between disease, poverty and urban space, what might be termed as spatial inequality.[57] In Britain, for example, premature death rates at all ages are two to three times higher among disadvantaged social groups than their more affluent counterparts. Contributory factors include the quality of housing, working conditions and pollution; economic and social influences, such as income and wealth; level of unemployment; quality of social relationships and social support; and access to effective health and social services.[58] The impact of the physical environment itself has been the subject of a recent review and continues to be researched widely, including its impact on levels of obesity due to poor access to healthy food or due to the quality of a person's local environment in encouraging walking and cycling.[59] There are still many unanswered questions regarding the relationship between health and space.

An early example of contemporary analysis of spatial inequalities is the work of the radical geographer William Bunge, whose *Nuclear War Atlas* from 1988 highlighted the disproportionate rates of injury amongst inhabitants of a slum area of Detroit (see Figure 2.17). Bunge mapped data on rates of injury from automobile accidents that he had collected in the 1960s, showing how they occurred on parts of the historical street network that were being used by commuters as a cut-through. The conflict between automobiles and pedestrians walking to school was, he argued, an entirely predictable pattern, one that was therefore not accidental at all. An earlier version of the map was more bluntly entitled *Where Commuters Run over Black Children on the Pointes-Downtown Track*.[60]

Bunge's blaming of the geography of Detroit's streets, namely the inner-city gridiron pattern, was a common critique of older 'small lot' cities, which in the past had been shaped for a scale of distance that could be travelled by horse and buggy. This supposedly anachronistic planning scale was one of several reasons given for the building of clean, safe

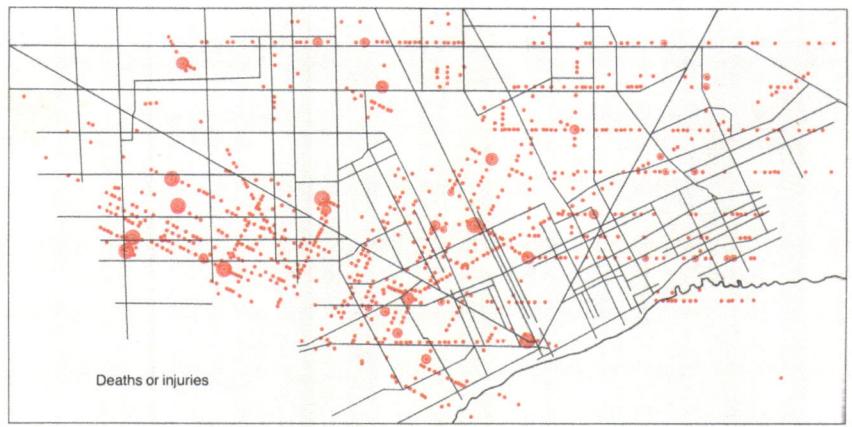

Map 2.16 Children's automobile 'accidents' in Detroit
Going to school forces children to cross dangerous streets. In front of schools as many as five or six 'accidents' occur like clockwork each year. If you can predict an event, why call it an 'accident'? If you can point to a corner and say 'next year five more kids will be hit here', it is the geography of the streets, the inner-city gridiron pattern left over from horse-and-buggy days, not negligent mothers, that is causing the deaths and injuries. If the streets were cul-de-sacs, as in the suburbs, these 'accidents' would all but cease.
Source: Detroit Police Department.

Figure 2.17 *Children's Automobile 'Accidents' in Detroit*, 1970.
W. Bunge, *Nuclear War Atlas* (Oxford: Basil Blackwell, 1988).

suburbs as a solution to inner-city problems. Putting aside the lack of evidence that through streets were more prone to pedestrian injuries from cars, Bunge's analysis highlights a much more complex situation: in impoverished neighbourhoods, which would at this time have had a disproportionately high African-American population, a wide variety of factors contributed to the accident rate, including a lack of investment in safe crossing or traffic calming measures. Even if dispersal to the suburbs was the correct solution, many inner-city residents would have been excluded from moving there, whether for reasons of affordability, prejudice or both (see section on redlining in Chapter 5).

Another more recent study from the US concerning spatial inequalities has suggested a multivariate connection between spatial morphology, housing quality and race. In a study of excess mortality amongst the residents of a pair of neighbourhoods in Chicago, 1995, Eric Klinenberg found a stark difference between survival rates in two adjacent neighbourhoods, North Lawndale and Little Village.[61] The two had ostensibly similar demographics, with large elderly populations living on their own, but North Lawndale had a death rate 10 times that of its neighbour. His analysis considered (amongst many factors) whether a variety of economic and social transformations had resulted in a sharp difference in the social ecology of the two areas. While North Lawndale was full of empty lots and abandoned buildings, Little Village had a dense concentration of buildings fronting the street, generating a network of social ties between business owners, customers and passers-by. This neighbourhood had a much higher residential population, with deeper kinship networks than North Lawndale. In contrast, North Lawndale had a low population density, and few social ties within or outside of the neighbourhood. This meant that when a heatwave struck, given the widespread lack of home air conditioning (and the expense of running it where it was present), North Lawndale residents did not seek refuge in local shops and businesses or amongst neighbours, and remained trapped in their overheated apartments: meanwhile 'the collective life of the area, the material substratum of busy streets, dense residential concentration, proximate family habitation, and booming commerce in Little Village fosters public activity and informal social support amongst the residents'.[62] Klinenberg points out that other factors influenced the death rate overall, such as power outages and a shortage of ambulances. However, these do not explain the local differences in mortality. David Seamon has pointed out that Klinenberg's analysis is akin to the view famously put forward by Jane Jacobs on how the urban layout is vital in shaping social interaction.[63] She argued that simple devices, such as street-facing buildings in a dense mesh of mixed uses, create opportunities for informal social interaction to take place and, over time, for casual acquaintanceship to develop.[64] In effect, Seamon argues, the differences between the two adjacent neighbourhoods' survival rate had a fundamental environmental explanation.

A similar pattern of spatial inequalities can be observed in the rates of excess mortality during the 2003 heatwaves in Paris (with a total of 15,000 deaths in three weeks). One study has found a disproportionate number of heat-related deaths amongst older people was due to high mean minimum nocturnal temperatures (namely, not enough of a relief from the heat at night)[65] that correlated with spatial and environmental factors such as surface temperature, vegetation and building materials. Interestingly, in a different study, Richard Keller has found that the Parisian cholera epidemic of 1849 and the pattern of deaths during the 2003 Paris heatwave are closely associated spatially.[66] Vulnerabilities, he writes, can build up over decades, the outcome of urban planning along with a political culture and social structure. In this instance, the large numbers of elderly women who died in the heatwave can be associated with the fraying of social networks for people living alone. Here, social isolation was concentrated in a number of inner-city districts in which there is a concentration of buildings dating back to Hausmann's reconfiguring of the Paris streetscape in the mid-nineteenth century. The city's existing network of narrow streets was said to inhibit circulation as well as contributing to miasmatic conditions (Figure 2.18). Haussmann's vast planning project, which continued for almost two decades from 1852, included the demolition of many of the narrowest streets in the city, creating a network of wide boulevards which cut through some of the city's poorest districts. The result was a relocation of many of the city's poor to its periphery, while those that remained found themselves situated in overcrowded districts with boundaries hardened by the grand boulevards. Keller's analysis of the 2003 heatwave found that the districts which Villermé had found in 1849 to be disproportionately poor, and suffering from excess deaths, matched the districts which had the greatest excess mortality in 2003.[67]

In addition to urban spatial conditions, Keller's analysis found that the architectural characteristics of the buildings dating back to Haussmann's times, which had tiny *chambres de bonne* (maids' quarters) at the top of the buildings, were disproportionately occupied by the city's elderly. Poorly ventilated, situated at the top of steep staircases, these were the least ideal locations for the frailest population of the district (see Figure 2.19). In essence, a form of vertical segregation had brought social exclusion down to the scale of buildings.

Just as in Paris, London has also shown signs of long-term persistence in spatial inequalities. In a recent study Danny Dorling, Scott Orford and colleagues looked at patterns of mortality from poverty-related illness, using data from the Charles Booth maps of poverty 1889 to compare the geography of poverty in the late nineteenth century with that of the late twentieth century. They concluded that the spatial persistence of poverty over time was extremely robust, despite a century of urban change having taken place since Booth's time.[68] Importantly, the mapping process aimed to go beyond the simple pursuit of visible patterns on the map – the authors tested for underlying spatial explanations for the measurable differences in health outcomes in different areas of a city.[69]

Another aspect of spatial inequality is the relationship between the location of pollution and respiratory disease. Alan Penn, Ben Croxford and

Figure 2.18 Rue Estienne, Paris, c. 1853. The buildings that lined the street were already being torn down when the photograph was taken. They disappeared to make way for the Rue du Pont-Neuf during Haussmann's renovation of Paris.

Photographer: Charles Marville. Photo from collections of the State Library of Victoria under the Accession Number: H88.19/29. Public domain.

colleagues have used space syntax methods to analyse the street grid configuration of an area of London to compare the predicted rates of vehicular flows and average and extreme CO concentrations, finding a strong relationship between the two, meaning that planners can take account of potential problems with pollution at the finest resolution of

Figure 2.19 Rooftop *chambres de bonne* in Paris, 2008.
Photograph by Rafael Garcia-Suarez via Wikimedia Commons.

single streets (whereas typically pollution monitoring will take account of average data covering a relatively large area, and without consideration of the street layout). More recent research that builds in the prevailing wind direction may also bear fruit in the future, while the increasing availability of cheap pollution monitors means that local communities are getting involved with citizen science projects to monitor, and provide evidence for, local pollution problems.[70]

The geography of health in relation to obesity (and the diseases associated with it, such as diabetes and heart disease) is one of the dominant topics in public health research in developed countries today, due to its prevalence as an issue of concern. A variety of indices that aim to model how much an environment encourages physical activity are used by health researchers to predict patterns of walking and/or cycling. A typical walkability index will be a combined model of residential and commercial density, land use mix and junction density. By quantifying how walkable an urban landscape is, health researchers can test whether environments have measurable effects on people's walking activity or on health outcomes (such as weight or blood pressure).[71] One such index developed for the United Kingdom incorporates space syntax measures to predict walking in London.[72]

In general, disease mapping today is both graphic and statistical: first describing the nature of the disease spatially, then attempting to determine its cause by hypothesising cause and effect. The map is only the starting point in the process of research investigation. Despite the fact that the field of research has advanced enormously since the first maps of yellow fever were plotted over two centuries ago, a lot still remains to be understood about the role of the built environment in health outcomes. While contemporary maps continue to advance in their sophistication, it is clear that further work needs to be done to understand this complex interrelationship. We know, for example, that health declines with poverty, and that poverty corresponds to certain types of spatial settings, but how these factors interrelate is not yet clear. The next chapter picks up the subject of poverty, taking us back to 1850s Liverpool.

Notes

1. 'The Downtown Back Alleys' in Riis, *How the Other Half Lives*, p. 33–4.
2. E.W. Gilbert, 'Pioneer Maps of Health and Disease in England,' *The Geographical Journal* 124, no. 2 (1958).
3. This is one of the earliest uses of the term *quarantine*. It relates to the Venetian dialect word for a unit of 40, *quarentena*, here denoting a period of 40 days. It is originally meant to relate to the period of Lent, though various other origins are given in the Oxford English Dictionary. In fact, the concept of spatially isolating disease dates to biblical times; e.g. 'All the days the lesion is upon him, he shall remain unclean . . . his dwelling shall be outside the camp'; Leviticus 13:46.
4. T. Koch, *Disease Maps: Epidemics on the Ground* (Chicago: University of Chicago Press, 2011), pp. 50 and 53.
5. For more on this conception of mapping as opposed to map-making, see Chapter 1 of T. Koch, *Cartographies of Disease: Maps, Mapping, and Medicine* (Redlands: California ESRI Press, 2017).
6. S. Griffiths, 'To Go with the Flows or to Flow with the Nodes? An Exploration of "Post-Disciplinary" Theories of Movement in Space Syntax and Mobilities Research,' in *11th International Space Syntax Symposium*, ed. T. Heitor (Chair), M. Serra, J.P. Silva, A. Tomé, M. B. Carreira, L.C. Da Silva and E. Bazaraite, 64.1–64.10 (Lisbon, Portugal: University of Lisbon, 2017), p. 64.6.
7. Quotation is from the Princeton University librarian's notes on Seaman's map from http://libweb5.princeton.edu/visual_materials/maps/websites/thematic-maps/quantitative/medicine/medicine.html. Accessed 18 July 2017. The plates in Figure 2.1 (and 1.2 earlier in this book) appeared in Seaman's 'An Inquiry into the Cause of the Prevalence of the Yellow Fever in New-York,' dated 10 March 1797.
8. Cosgrove, *Geography and Vision*.
9. Robinson, *Early Thematic Mapping in the History of Cartography*.
10. The map notes can be found in Princeton University Library, 'Medicine,' in *First X, Then Y, Now Z: Landmark Thematic Maps*, Princeton University Library Historic Maps Collection (2012), http://libweb5.princeton.edu/visual_materials/maps/websites/thematic-maps/quantitative/medicine/medicine.html (accessed 11 April 2018). Although he misattributed the cause of the disease, it is important to note his use of dots to record the locations of disease, some six years before John Snow.

11. T. Shapter, *The History of the Cholera in Exeter* (London: John Churchill, 1832).
12. See http://libweb5.princeton.edu/visual_materials/maps/websites/thematic-maps/quantitative/medicine/medicine.html.
13. H.W. Acland, *Memoir on the Cholera at Oxford, in the Year 1854, with Considerations Suggested by the Epidemic* (London: John Churchill, 1856), pp. 51, 57. The map can be viewed at http://libweb5.princeton.edu/visual_materials/maps/websites/thematic-maps/quantitative/medicine/medicine.html.
14. See S. Schulten, *Mapping the Nation: History and Cartography in Nineteenth-Century America* (Chicago: University of Chicago Press, 2012), p. 70 and electronic version of the map at: http://mappingthenation.com/index.php/viewer/index/3/6. The map accompanied the report of the Committee of Internal Health on the Asiatic Cholera, together with a report of the city physician on the Cholera Hospital (Boston, MA: J.H. Eastburn, 1849).
15. The map's legend states that 'less cleansed districts are marked in dark brown'. Chadwick's report mentions house owners being brought before a magistrate for neglecting to cleanse and whitewash their property between tenants, this being especially bad in the poorer districts. 'Even the best streets are very badly cleansed, but in the poorer streets of the city the cleansing is very bad indeed — horribly bad. Take Duke's Place, for example; you will see cabbage-stalks and rotten oranges that have been thrown away, and they often remain there for several days. We do not get our streets swept oftener than once a-week.' E. Chadwick, 'Report on the Sanitary Conditions of the Labouring Population of Great Britain,' in Report to Her Majesty's Principal Secretary of State for the Home Department, from the Poor Law Commissioners, on an Inquiry into the Sanitary Condition of the Labouring Population of Great Britain (London: W. Clowes and Sons, 1842), p. 225.
16. 'A navigable river, canals . . . communicating with the Mersey at Liverpool . . . and thence with the Humber . . . railways branching off in every direction . . . These advantages give every possible facility for bringing raw materials, sending away manufactured goods, and for the access of men of business'. Quote from A. Kershen, *Uniting the Tailors: Trade Unionism Amongst the Tailors of London and Leeds, 1870–1939* (Ilford, Essex: Frank Cass & Co, 1995), p. 25.
17. Chadwick, *Report*, p. 160. Mr Baker was a poor-law surgeon in Leeds, namely, he had responsibility under the terms of the Poor Law for examining and certifying the conditions of workhouse inmates and the poor of the district. His report from 1833 was the source cited by Chadwick in his report.
18. Chadwick, *Report*, pp. 369–70, quoted in the map collection's notes. https://digital.library.cornell.edu/catalog/ss:19343540. Accessed July 21, 2017.
19. The original analysis of Leeds was published in L. Vaughan and A. Penn, 'Jewish Immigrant Settlement Patterns in Manchester and Leeds 1881,' *Urban Studies* 43, no. 3 (2006), and is re-examined here.
20. Quoted in *Leeds Mercury*, 16 June 1888.
21. The area also contained Kirkgate Market, which is where Mr Marks (subsequently of Marks and Spencer) opened his first 'penny bazaar'.
22. *Zu Rothenburg's Cholera-Epidemie des Jahres 1832 in Hamburg*. Lithograph map, with added colour. From J.N.C. Rothenburg, *Die Cholera-Epidemie des Jahres 1832 in Hamburg: Ein Vortrag, gehalten im der wissenschaftlichen Versammlung des ärztlichen Vereins, am 17 November 1835* (Hamburg: Perthes & Besser, 1836). The map can be viewed on the Princeton University Historic Maps Collection site: http://libweb5.princeton.edu/visual_materials/maps/websites/thematic-maps/quantitative/medicine/medicine.html.
23. E.-C. Hellis, *Memories of Cholera in 1832* (in French) (Paris: Ballière, 1833), held at the Bibliothèque Nationale de France, Department of Science and Technology. 8-TD57-389: http://catalogue.bnf.fr/ark:/12148/cb30589490z.
24. É. Daudé, E. Eliot, and E. Bonnet, 'Cholera in the 19th Century: Constructing Epidemiological Risk with Complexity Methodologies' (paper presented at the 3rd International Conference on Complex Systems and Applications, University of Le Havre, Normandy, France, 2009), especially page 6.
25. C. Dickens, *Nicholas Nickleby*, Chapter 2. University of Oxford Text Archive. http://ota.ox.ac.uk/text/3082.html. Accessed 21 March 2018.
26. See full analysis published in L. Vaughan, 'The Spatial Form of Poverty in Charles Booth's London,' *Progress in Planning: special issue on The Syntax of Segregation*, edited by Laura Vaughan 67, no. 3 (2007): 231–50.

27. R. Perry, *Facts and Observations on the Sanitory State of Glasgow During the Last Year: With Statistical Tables of the Late Epidemic, Shewing the Connection Existing between Poverty, Disease, and Crime* (Glasgow: Glasgow Royal Asylum for Lunatics, printed at the institution, 1844), p. 9, p. 22.
28. John Snow, *On the Mode of Communication of Cholera* (London: John Churchill, 1855). The book, as well as a treasure trove of maps from Snow's time and a broader literature on the use of mapping in epidemiology, can be found on the UCLA website at http://www.ph.ucla.edu/epi/snow.html. Another version of the map, prepared for the local Parish Enquiry Committee, has an additional dotted line to indicate the area within walking distance from the offending pump: see T. Koch and K. Denike, 'Essential, Illustrative, or . . . Just Propaganda? Rethinking John Snow's Broad Street Map,' *Cartographica: The International Journal for Geographic Information and Geovisualization* 45, no. 1 (2010), p. 23.
29. Gilbert, 2002, p. 20.
30. See analysis by Cliff and Haggett, cited in A. Barford and D. Dorling, 'Mapping Disease Patterns,' in *Wiley Statsref: Statistics Reference Online* (John Wiley & Sons, Ltd, 2014), p. 2, which shows the geographical centres of the epidemic alongside Snow's mapping.
31. W. Farr, 'Report on the Cholera Epidemic of 1866 in England: Supplement to the Twenty-Ninth Annual Report of the Registrar-General of Births, Deaths, and Marriages in England,' (London: HMSO, 1868), p. xi, cited in T. Cicak and N. Tynan, 'Mapping London's Water Companies and Cholera Deaths,' *The London Journal* 40, no. 1 (2015).
32. S. Johnson, *The Ghost Map* (London: Penguin Books, 2008), p. 199.
33. Koch and Denike, 'Essential, Illustrative, or . . . Just Propaganda?', p. 21.
34. Koch and Denike, 'Essential, Illustrative, or . . . Just Propaganda?', p. 28.
35. Gilbert, 2002, p. 13.
36. Gilbert, 2002, p. 24.
37. Gilbert, 2002, p. 25.
38. Booth, quoted in J.A. Yelling, *Slums and Slum Clearance in Victorian London*, vol. 10, The London Research Series in Geography (London: Allen and Unwin, 1986), p. 52. Yelling points out that Booth coined the term 'poverty trap' to describe streets being cut off from 'communication with the surrounding district'.
39. H.J. Dyos, 'The Slums of Victorian London,' *Victorian Studies* XI (1967), p. 25.
40. Redrawn from R. Plunz, *A History of Housing in New York* (New York: Columbia University Press, 1990); published in G. Palaiologou and L. Vaughan, 'The Sociability of the Street Interface – Revisiting West Village, Manhattan,' in *21st International Seminar on Urban Form - ISUF2014: Our Common Future in Urban Morphology*, ed. V. Oliveira et al. (Porto, Portugal: FEUP, 2014).
41. A.M. Blake, *How New York Became American, 1890–1924* (Johns Hopkins University Press, 2009), pp. 34–5. See also Plunz, *History of Housing in New York*.
42. Riis, *How the Other Half Lives*, p. 25. The Tenement House Committee maps of nationalities and population density from 1895 are coincidentally an important record of what Riis is describing.
43. No. 1. *Map of City of New York showing Densities of Population in the several Sanitary Districts, June 1, 1894*. No. 2. *Map of City of New York showing the Distribution of Principal Nationalities by Sanitary Districts*. Both made under the direction of the Tenement House Committee of 1894. https://digital.library.cornell.edu/catalog/ss:3293866. Created by Frederick Erastus Pierce (1878–1935). Source: *Harper's Weekly*, 19 January 1895.
44. J.B. Huber, *Consumption, Its Relation to Man and His Civilization, Its Prevention and Cure* (Philadelphia: Lippincott, 1906). The plan appeared on page 147. The publication date is only an estimate. It seems likely that it was published around the time it first appeared in the press in 1903.
45. Map image from Huber, *Consumption*.
46. Jacob Riis also describes buildings such as this, with their 'dirt and desolation' reigning in a tenement leading to 'a dark and nameless alley, shut in by high brick walls'. Riis, *How the Other Half Lives*, Chapter IV: 'The Down Town Back-Alleys', p. 149.
47. *The Brooklyn Daily Eagle*, 13 September 1903 (Sunday), p. 11.
48. *Los Angeles Herald*, Number 1, 2 October 1903.
49. C. Booth, 'Improved Means of Locomotion as a First Step Towards the Cure of the Housing Difficulties of London,' *Abstract of the Proceedings of Two Conferences Convened by Albert Browning Hall, Walworth* (London: Macmillan, 1901), p. 23. Sources have his contemporary

(General) William Booth, founder of the Salvation Army, advocating moving the poor from the filth and squalor of the slums to 'a neat little cottage in the pure air of the country'.
50. The map was published in T.J. Jones, *Directory of Inhabited Alleys of Washington* (Washington: Housing Committee Monday Evening Club, 1912), p. 1. The scan from the directory is courtesy Jenny Schrader from her web log http://jennysschrader.com/2017/01/alley-life-in-washington-dc-1920s/.
51. T.J. Jones, *Directory*, p. 6.
52. J. Borchert, *Alley Life in Washington: Family, Community, Religion, and Folklife in the City, 1850–1970* (University of Illinois Press, 1980).
53. See historical review in B. Clarke, 'Mapping the Methodologies of Burkitt Lymphoma,' *Studies in History and Philosophy of Science Part C: Studies in History and Philosophy of Biological and Biomedical Sciences* 48 (2014).
54. R.R. Frerichs, P.S. Keim, R. Barrais and R. Piarroux, 'Nepalese Origin of Cholera Epidemic in Haiti,' *Clinical Microbiology and Infection* 18, no. 6 (2012); R.R. Frerichs, *Deadly River: Cholera and Cover-up in Post-Earthquake Haiti* (Ithaca: Cornell University Press, 2016).
55. See review in Barford and Dorling, 'Mapping Disease Patterns', of the pitfalls in mapping the geography of disease.
56. Access to clean water is not necessarily a given even in the developed world today, as has been seen by the scandal of lead-contaminated water in Flint, Michigan. The location of Flint homes with >15 ppb of lead as of 1 February 2016 was published in the Detroit Free Press; see R. Allen, 'Flint Map: Where Lead Levels in Water Remain Too High,' *Detroit Free Press*, 2 February 2016, http://www.freep.com/story/news/local/michigan/flint-water-crisis/2016/02/02/flint-lead-map/79686158/. Here also the affected population is disproportionately from impoverished areas of the city.
57. See for example D. Acheson, 'Report of the Independent Inquiry into Inequalities in Health,' (London: Stationery Office; Department of Health, 1998); Low Income Project Team for the Nutrition Task Force, 'Low Income, Food, Nutrition, and Health: Strategies for Improvement,' (London: Department of Health, 1996).
58. Social interaction and urban design research dates back to the influential early work of Appleyard and Lintel, who found that the amount of vehicular traffic on streets had an impact on the ability of people living on them to form friendship ties. It has been revised recently by Jennifer Mindell and colleagues, who have looked at the impact of community severance – namely the effect of busy roads on people's access to goods and services and the consequent impact this has on health (whether people's willingness to walk, or their ability to be socially engaged: a lack of social engagement in itself can have a negative impact on health). J. Mindell and S. Karlsen, 'Community Severance and Health: What Do We Actually Know?,' *Journal of Urban Health* (2012); J. Mindell et al., 'Using Triangulation to Assess a Suite of Tools to Measure Community Severance,' *Journal of Transport Geography* 60, no. April 2017 (2017).
59. I. Geddes et al., 'The Marmot Review: Implications for Spatial Planning,' (London: Institute of Health Equity, UCL, 2011).
60. Information on the map is from the 'Collector's Notes' on the map on the Cornell University website, https://digital.library.cornell.edu/catalog/ss:19343514. A sister map to this one shows the point location of 'Region of Rat-Bitten Babies' in the same district and aims to demonstrate that virtually all 'frequent rat sightings' in Detroit – and every one of the 'Confirmed rat bites, 1967, 1969 and 1970' – occurred in the inner-city Fitzgerald 'slum ghetto'. See notes on this map at the Cornell University website https://digital.library.cornell.edu/catalog/ss:19343517.
61. E. Klinenberg, *Heat Wave: A Social Autopsy of Disaster in Chicago* (Chicago: University of Chicago Press, 2005).
62. Klinenberg, *Heat Wave*, p. 109.
63. D. Seamon, 'Lived Bodies, Place, and Phenomenology: Insights from Edmund Husserl, Maurice Merleau-Ponty, Edward Casey, Jane Jacobs, and Eric Klinenberg,' *Journal of Human Rights and the Environment* 4, no. 2 (special issue on human bodies and material space) (2013).
64. J. Jacobs, *The Death and Life of Great American Cities* (Harmondsworth: Penguin, 1961).
65. B. Dousset et al., 'Satellite Monitoring of Summer Heat Waves in the Paris Metropolitan Area,' *International Journal of Climatology* 31, no. 2 (2011), p. 321.
66. F. Canoui-Poitrine, E. Cadot and A. Spira, 'Excess Deaths During the August 2003 Heat Wave in Paris, France,' *Revue d'Épidémiologie et de Santé Publique* 54, no. 2 (2006). The authors plotted

excess mortality (namely where the number of deaths exceeds what would be expected within a population) across the city to see if there is a relationship between spatial location and excess mortality, taking account of factors such as socioeconomic status and population age in each neighbourhood.
67. R.C. Keller, *Fatal Isolation: The Devastating Paris Heat Wave of 2003* (Chicago: University of Chicago Press, 2015), p. 101. Interestingly, Felicity Edholm recounts how, by the 1880s, 'working-class women spent most of their time within a quite tightly defined local area and would, within this area, be part of a community of women'. F. Edholm, 'The View from Below: Paris in the 1880s,' in *Landscape: Politics and Perspectives*, ed. B. Bender, Explorations in Anthropology: A University College London Series (Oxford: Berg Publishers, 1995), pp. 159–60.
68. D. Dorling et al., 'The Ghost of Christmas Past: Health Effects of Poverty in London in 1896 and 1991,' *British Medical Journal* 321 (2000); S. Orford et al., 'Life and Death of the People of London: A Historical GIS of Charles Booth's Inquiry,' *Health and Place* 8, no. 1 (GIS Special Issue) (2002). See also L. Vaughan and I. Geddes, 'Urban Form and Deprivation: A Contemporary Proxy for Charles Booth's Analysis of Poverty,' *Radical Statistics* 99 (2009), a discussion of which is elaborated in Chapter 4.
69. The Booth map has also been used to compare data on diabetes in an area of East London to see if there are associations with poverty pockets persisting from the past. See D. Noble et al., 'Feasibility Study of Geospatial Mapping of Chronic Disease Risk to Inform Public Health Commissioning,' *BMJ Open* 2, no. 1 (2012), and their images at http://news.bbc.co.uk/1/shared/bsp/hi/pdfs/12345.pdf.
70. B. Croxford, A. Penn and B. Hillier, 'Spatial Distribution of Urban Pollution: Civilizing Urban Traffic,' *Science of The Total Environment* 189 (1996); A. Penn and B. Croxford, 'Effects of Street Grid Configuration on Kerbside Concentrations of Vehicular Emissions,' in *1st International Space Syntax Symposium*, ed. Major, M. D., L. Amorim and F. Dufaux, 27.1–27.10. London: University College London, 1997). The results were ($r^2=.78$, $p=.0002$). Recent citizen science projects of this nature have been carried out by a London-based social enterprise, Mapping for Change.
71. For example M.P. Buman et al., 'Objective Light-Intensity Physical Activity Associations with Rated Health in Older Adults,' *American Journal of Epidemiology* 172, no. 10 (2010).
72. D. Van Dyck et al., 'Perceived Neighborhood Environmental Attributes Associated with Adults' Transport-Related Walking and Cycling: Findings from the USA, Australia and Belgium,' *International Journal of Behavioral Nutrition and Physical Activity* 9, no. 70 (2012). An alternative index has been developed for the United Kingdom, incorporating space syntax measures to predict walking in London: see A. Dhanani, L. Tarkhanyan and L. Vaughan, 'Estimating Pedestrian Demand for Active Transport Evaluation and Planning,' *Transportation Research Part A: Policy and Practice* 103 (2017).

3
Charles Booth and the mapping of poverty

From where, off Shoreditch High Street, a narrow passage, set across with posts, gave menacing entrance on one end of Old Jago Street, to where the other end lost itself in the black beyond Jago Row; from where Jago Row began south at Meakin Street, to where it ended north at Honey Lane – there the Jago, for one hundred years the blackest pit in London, lay and festered; and half-way along Old Jago Street a narrow archway gave upon Jago Court, the blackest hole in all that pit.[1]

Social conditions: observing the problem

Before we continue further into these two centuries' worth of cartographic wanderings, it is important to discuss the influence of the work of the great social reformer and father of social investigation, Charles Booth.

Booth is not an unknown figure in the social sciences. At least 10 books have been dedicated to establishing his importance as the father of the field, while the two series of maps which accompanied Booth's 17-volume, 14-year project are themselves famous for their cartographic importance. Yet, possibly due to the ongoing division between the social and the spatial sciences, few of the books about Booth written in the past quarter century devote more than a passing mention to the maps.[2]

This is the first of this book's two chapters dealing with poverty maps. It will lay out the essential importance of Booth's maps, which have been described as an 'elaborate exercise in social topography',[3] and what they tell us about the relationship between urban configuration and poverty. The second chapter will show the impact Booth had on a small number of other early social scientists either side of the Atlantic, from which we will gather more about the spatial form of poverty in Chicago, after which we will turn to York and then back to London.

We will start with a divergence from the conventional history of social cartography, which places Charles Booth as the progenitor of the poverty map. A significant precursor to his work is found in the set of maps of Liverpool published in 1858 by a parish priest, Abraham Hume, who lived in All Souls, Vauxhall – possibly the poorest and unhealthiest district of the poorest city in England (see Figure 3.1).

Maps were a constant in Hume's method. Prior to the Liverpool study, he had used them for his contribution to the 1851 census of religious worship (which was the first and only time until recent history when an enquiry into religion formed part of the official decennial census). The census used four different schedules to record attendance at the established and non-established churches of England and Wales and of Scotland. Hume subsequently gave evidence before two select

Figure 3.1 Rev. Abraham Hume's *Map of Liverpool, Ecclesiastical and Social* (coloured historically), 1854.

From A. Hume, *Condition of Liverpool, Religious and Social, Etc.* 2nd ed. (Liverpool: Privately printed, 1858). Electronic image courtesy of Tinho da Cruz, Department of Geography and Planning, University of Liverpool.

committees of the House of Lords on 'Means of Divine Worship in Populous Places' and 'Church Rates', and went on to conduct many other religious geographical studies in later years.

Deeply concerned by the results of the religious worship census, which seemed to suggest that the significant reduction in church-going and a surprisingly high non-conformist presence meant the country was on its way to being 'a heathen society', Hume undertook to investigate the situation locally, to see if church-going was associated with poverty. Almost as soon as he had been given charge of his Liverpool parish Hume organised six theological students to visit the 2,400 houses in the area and compile records on their housing conditions as well as patterns of church-going. Hume set out his statistics street by street in a pamphlet published in 1858, alongside a set of maps – which were in fact the same map coloured up four different ways – to denote, in turn, ecclesiastical, historical, municipal and moral and social statistics.[4]

Hume explained his method as follows.

> Desirous of ascertaining it more minutely, I applied to all the relieving officers within the Borough; and they very kindly furnished me with lists of streets in their respective districts, in which outdoor relief is most uniformly distributed. Each of them divided his list of streets into two classes; those which were wholly pauper, and those which were half or partially so. There were nearly 200 streets included in all; viz – 56 of the former kind, and 139 of the latter. All of these are indicated on the map, by dark serrated marks, which are denser in the former than in the latter.[5]

Hume's map was marked up with dark serrated marks for 'pauper' streets and light serrated marks (in reality, more sparsely drawn) for 'semi-pauper' streets as well as chapels and churches and related ecclesiastical data. His data on the location of streets of crime and immorality, and the location of poverty streets (divided into semi-pauper streets and pauper streets), took up a considerable amount of the text in his pamphlet, where he also discussed the relationship between church-going and poverty. His notation of areas of cholera and violent deaths in an inset of the map indicates his association of poverty with disease on the one hand and crime on the other.

We can also see a section of the poorest district in the section of the map reproduced in Figure 3.2, which is centred on Hume's own parish area and the map's key is in Figure 3.3.

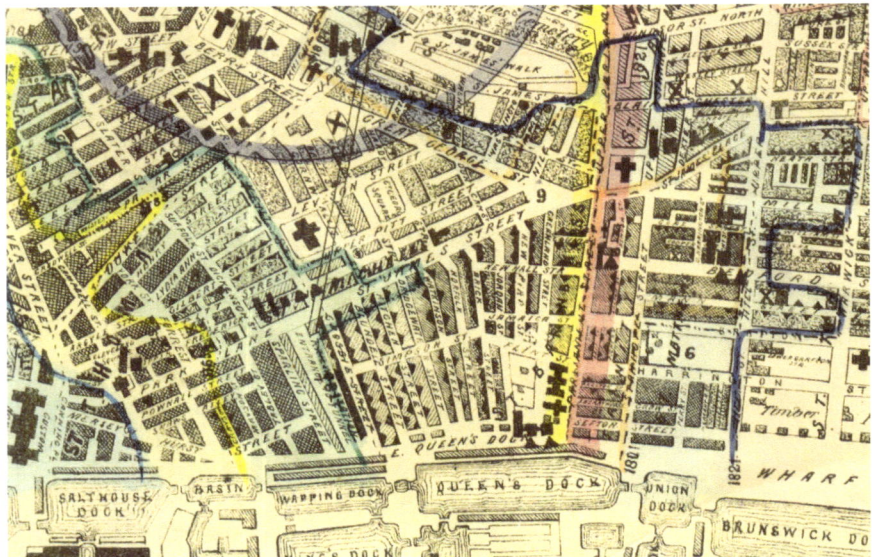

Figure 3.2 Detail of: Rev. Abraham Hume's *Map of Liverpool, Ecclesiastical and Social*
From A. Hume, *Condition of Liverpool, Religious and Social, Etc.* 2nd ed. (Liverpool: Privately printed, 1858).

Although this was a survey that focused on religious observance, Hume's descriptions of the local conditions point to how attuned he was to the location of poverty in the interstices of the district:

> the moment we diverge from these main lines to the by-ways which are less known, we find destitution of every degree, and crime and suffering of every kind. All the lower part of Toxteth Park, lying along the line of the river but at some distance, is of this kind; and Everton which within the last thirty years was an elegant suburban retreat, is now crowded with an humble population. These broad distinctions will be sufficient to indicate the 'region of pauperism'.[6]

Looking at the map it is clear that the poor were living in tight clusters in a small number of streets. Many of these would have been back-to-back dwellings that the Liverpool Sanitary Act of 1842 (possibly the first public health legislation in England) had condemned a few years earlier, due to the lack of ventilation and sanitation within them (see Figure 3.4).[7] Hume finds the largest clustering of pauperism to be located in St. Thomas', Toxteth, and the next in order in St. James'. Overall, he finds poverty in 195 streets, but smaller localities have

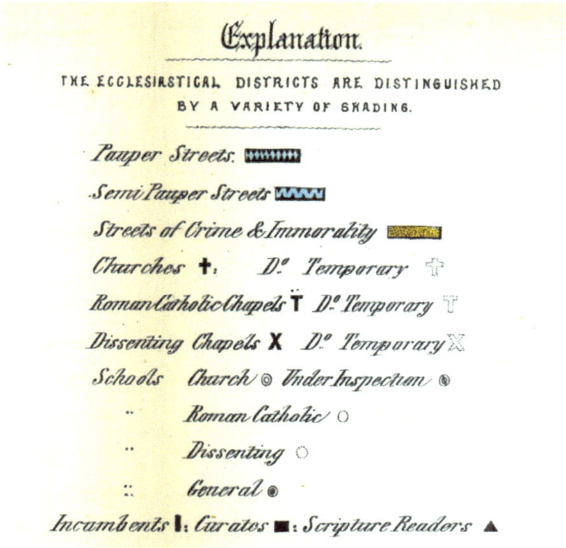

Figure 3.3 Key to Rev. Abraham Hume's *Map of Liverpool, Ecclesiastical and Social*

From A. Hume, *Condition of Liverpool, Religious and Social, Etc.* 2nd ed. (Liverpool: Privately printed, 1858).

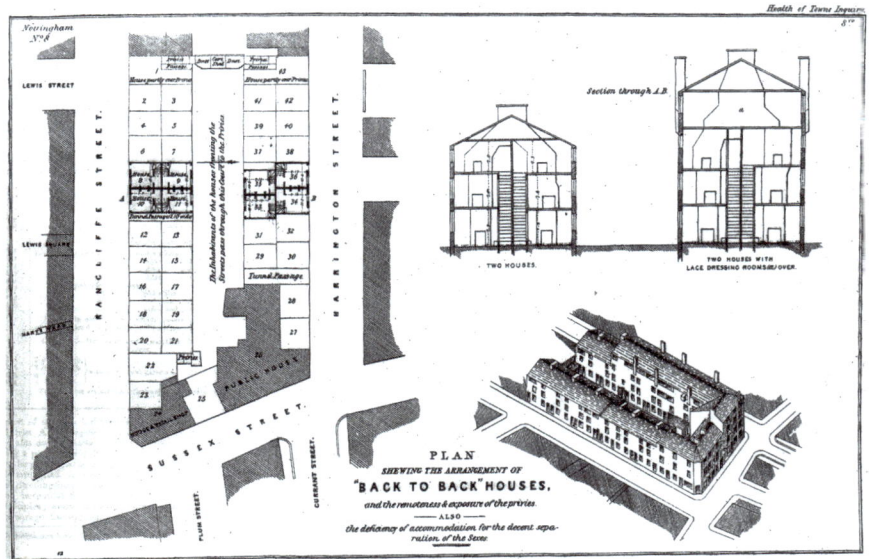

Figure 3.4 Thomas Hawksley's *Plan Shewing the Arrangement of 'Back to Back' Houses and the Remoteness and Exposure of the Privies, also the Deficiency of Accommodation for the Decent Separation of the Sexes.*

T. Hawksley, 1844. Wellcome Collection, https://wellcomecollection.org/works/mcwh8yh4.

clusters – 'specially devoted' to crime, vice and immorality – in 'only' 33 streets in total.[8]

The city's growing prosperity meant that the middle classes could increasingly afford to move away from the overcrowded inner city, so their presence in the centre diminished. Hume argued that not only had their presence provided a positive influence, but the middle classes had also cushioned the worst of the poverty through their contributions to local church taxes and support for community amenities such as libraries, schools and hospitals; in parallel, the greater demand on day labour at the nearby docks resulted in the creation of 'an exclusive belt surrounding the docks in which the density of population was greater than anywhere else in the city'.[9] Hume points out how the city's topography helped increase overcrowding, which he saw as an important contribution to poverty, since suburbanisation was constrained by the presence of the docks to the west. The result was the formation of a close-packed band of poverty by the docks, surrounded at a distance away by a semi-circle of more prosperous households.

Space syntax analysis of Liverpool's urban morphological evolution from the 1850s to today has shown how the city's original arrangement of radial streets emanating out from the docks created a remarkably enduring pattern of localised spatial segregation in areas such as the one here, with Princes Avenue (labelled New Road on the map's right-hand side) demarcating the district of Toxteth-Granby that, after a spell as a relatively prosperous suburban enclave of Jewish immigration, went into a deep decline.[10] This decline was partially due to the collapse in the city's shipping industry in the 1930s along with the widespread damage wrought by the heavy bombing of the docks (which were an important conduit of supplies during the Second World War). An earlier study of Toxteth, in the post-war period, shows how a combination of severe economic deprivation along with the ongoing spatial and social isolation of the district's minority black community had led by that time to the local population being significantly disadvantaged, with the perception of Toxteth as a 'no-go' area only exacerbated by the 1981 riots that took place in the district. The authors pointed to the need for any social regeneration to take account of the lack of mixing as well as the lack of wider spatial connectivity, to ensure any improvement to the local community's situation.[11] Indeed it has taken until the turn of the new millennium for the area to start to benefit from the general uplift in the city's circumstances, coupled with targeted social projects.

Social geography: diagnosing the problem

Two thousand copies of Hume's map were sold or distributed. One was set before the select committee of the House of Lords, *Appointed to Inquire into the Deficiency of Means of Spiritual Instruction and Places of Divine Worship in the Metropolis, and in Other Populous Districts in England and Wales, Especially in the Mining and Manufacturing Districts* (and etc.) of 1857–8; the map was also displayed at the National Association for the Promotion of Social Sciences meeting held in the same year in Liverpool. Only a few years later, Charles Booth, a shipping industrialist then still living in Liverpool, was on the campaign trail in the city for a Liberal Party seat for the national elections of 1865. His diaries show how shocked he was to see the extent of poverty in Toxteth when canvassing in the area. While there is only circumstantial evidence for him having seen Hume's maps, it seems more than a coincidence that when Booth embarked on his survey in 1880s London he employed similar methods of gathering statistics and mapping them to those that had been used by Hume. As David Smith has noted, Hume's maps could very well be considered a precursor to Booth's.[12] Nevertheless, in scale, ambition and influence, the two men's efforts cannot compare.

Although, like Hume, Booth was not a social theorist, he was an empiricist. His approach to studying the conditions underlying poverty was based on a confidence that the science of statistics was fundamental to social progress. Having staked a considerable fortune from his shipping business, he embarked on the study that became his life's work.

Aiming to understand the causes and contributory factors to poverty, Booth planned his research in a business-like way. This was especially important in a context where other theories were lacking a factual basis; as his biographers wrote: 'no one really knew the truth about how the poor lived. It was if they were living behind a curtain on which were painted terrible pictures.'[13]

What is particularly interesting is the social milieu in which Booth undertook his project. The late nineteenth century was a time of growing concern about the very nature of urban society. As Reeder states in his introduction to the reproduction of the 1889 poverty maps,

> during the 1880s a new perception was being formed of London's social condition, growing out of a spate of writings on how the poor lived by journalists and city missionaries . . . Middle-class anxieties were fuelled by descriptions of . . . the poor as a brutalised and

degenerate race of people, the victims but also the agents of the deteriorating forces in city life.[14]

Booth's work coincided with a critical point in the social and economic history of London. This was a period when London's labour market was based on small-scale production and the finishing trades. Work was seasonal, and workers were employed on a casual basis, with employment rates fluctuating with the demands of the market. Few were experts in a single trade. Instead, workers would hold several occupations throughout a single year (a fact that would also need to be considered when examining census records). Anna Davin shows how this pattern of work made workers much more spatially dependent; knowledge of casual work, or references for work or charity, were reliant on local knowledge built up through long-standing residence in the area. This had the greatest detrimental effect on the weakest and least powerful people socially; 'those who depended on their local environment the most to support them in their everyday life'.[15] Yet the poorest classes were most likely to have to move frequently due to changes in income or rent costs. This enforced transience increased their relative disadvantage: 'variations in family income or household composition were often a reason for changing house. When income shrank through illness or unemployment, leaving even less margin for rent, somewhere cheaper had to be found.'[16] Davin shows that this situation of supposed restlessness was criticised by the so-called comfortable classes, who had little contact with the poor, except through the accounts of reformers and professionals such as clergy and public health inspectors. Paradoxically, the poor were also criticised for being immobile, concentrating in large masses of disease and immorality.

The degradation of the physical environment in poverty areas became a matter of increasing public concern, with many campaigners writing pamphlets on the subject. One of the most influential was that by Reverend Andrew Mearns, *The Bitter Cry of Outcast London*, which resonated widely in its descriptions of the 'pestilential human rookeries' in which the poor were living:

> We do not say the condition of their homes, for how can those places be called homes . . . Few who will read these pages have any conception of what these pestilential human rookeries are, where tens of thousands are crowded together . . . To get into them you have to penetrate courts reeking with poisonous and malodorous gases arising from accumulations of sewage and refuse scattered in all directions and often flowing beneath your feet.[17]

The poor problem had become a spatial problem, since high concentrations of poverty were seen to risk the moral and physical contamination by the casual poor of the respectable poor. Booth's enquiry, then, started in a context wherein poverty was considered a moral problem. As we saw in Hume's study, a correlation between an absence of church-going, excess of drink and general moral imprudence was associated in the popular imagination as the root of the problem of poverty. Political concerns with this matter came to a peak with the 1886 Trafalgar Square riots, which brought to public attention the presence of a mass of poverty within shouting distance of London's heartland. Six months later Charles Booth started his survey, with his team setting out on foot to conduct a preliminary inquiry into the poverty conditions and occupations of the people of East London. This study aimed to classify every street, court and block of buildings across the metropolis by direct observation, but it soon became clear that in order to cover the entire city Booth's team would need to switch to capturing statistics on a street-by-street basis.

Booth's city-wide enquiry essentially aimed to discover how many people were living in poverty, what kept them in this state and what might be done to alleviate it. His ability to force the 'poverty question' away from a debate about morality and towards a consideration of practical solutions was a significant one: he showed that poverty was more likely to be due to under- or unemployment and less likely due to personal failure (drunkenness and the like). Importantly, Booth's classes were based on income combined with employment patterns and status, rather than social classes. Once Booth had compiled his statistics he consulted experts on the accuracy of his coding: in fact, Booth's ambition to be statistically rigorous has been validated in recent years, with research into his classificatory scheme showing it to be internally consistent in how it differentiates the poverty classes.[18]

In one of his first addresses about the survey to the Royal Statistical Society Booth stated: 'it is the sense of helplessness that tries everyone; the wage earners, as I have said, are helpless to regulate or obtain the value of their work . . . We need to begin with a true picture of the modern industrial organism . . . it is the possibility of such a picture as this that I wish to suggest.'[19] By this time Booth was becoming aware of current thinking in the fields of social reform and practical philanthropy. Ruskin, Octavia Hill and others were talking about giving time and intelligence to finding a solution to poverty in preference to simply making charitable donations.[20] This thinking was also in tune with Booth's ongoing conversations with his future collaborator, Beatrice Webb, about the possibilities of social diagnosis through scientific enquiry.[21] At the same time there was widespread publicity regarding the conclusions from Henry Hyndman's enquiry for the Social Democratic Federation, published in 1885, which claimed that 25 per cent of the population lived in conditions of extreme poverty. It is said that this claim helped bolster Booth's resolve to conduct a study that would disprove what seemed to be an unlikely statistic.[22]

In addition to his study of poverty, Booth carried out a series of detailed studies on the working conditions in the principal London industries, as well as on specific subjects he felt relevant to the study of life and labour, such as the 'sweated' industries. His series ended with an inquiry into 'Religious Influences', which included detailed accounts of his interviews with clergy from across the religious spectrum.

Booth's preliminary results were shocking. While Hyndman's estimate of 25 per cent was seen to be excessive, Booth found 33 per cent of London living in poverty, with an even greater proportion, 35 per cent, among those living in the city's East End. Not only had he determined that the rate was greater than had been estimated previously, his detailed reports shed much light on the nature of poverty. One of his many important findings was that regularity of income was as significant to poverty as its level. Booth showed that people working in certain industries, such as tailoring, which suffered from ebbs and flows throughout the year, experienced similar fluctuations in rates of poverty, such that they were never assured of a steady income. He also showed that poverty was linked to where people lived.

The street study, which started in 1889, was captured on a detailed scale map (6 inches to 1 mile or 1: 10,560) according to finely delineated gradations of poverty and prosperity from black, dark blue and light blue for the poverty classes through pink and red to gold for the wealthiest. From a cartographic point of view, one of the striking aspects of the study is that the unit of analysis was the street segment, namely the section of street between two junctions, so the finest variations could be recorded, occasionally even differentiating different sides of the same street. Yet it is likely the darker colours would have been chosen to meld together from afar to emphasise clusters of poverty streets. The colour scheme used on the map was as follows:

> Black. The lowest grade (corresponding to Class A in the statistical study), inhabited principally by occasional laborers, loafers, and semi-criminals – the elements of disorder.
>
> Dark Blue. Very poor (corresponding to Class B), inhabited principally by casual labourers and others living from hand to mouth.

Light Blue. Standard poverty (corresponding to Classes C and D) inhabited principally by those whose earnings are small... whether they are so because of irregularity of work (C) or because of a low rate of pay (D).

Purple. Mixed with poverty (usually C and D with E and F, but including Class B in many cases).[23]

Pink. Working-class comfort. Corresponding to Class E and F, but containing also a large proportion of the lower middle class of small tradesman and Class G. These people keep no servants.

Red. Well-to-do; inhabited by middle-class families who keep one or two servants.

Yellow. Wealthy; hardly found in East London and little found in South London; inhabited by families who keep three or more servants, and whose houses are rated at £100 or more.

The data used for these classifications were based on several sources, first and foremost among which were the School Board Visitors, who had a detailed knowledge of families with children. This was a well-judged decision, as their role (to ensure school attendance) required familiarity with the entire family and its living conditions, as those who were living in the worst conditions did not have to make even the minimal payment for school. As part of their regular round of inspections, which sometimes would have been repeated over several years, such visitors would record the state of the housing, of its inhabitants and whether the breadwinners of the household were fit to work, in addition to the size of the household and – importantly – the regularity of its income. The visitors' information was cross-checked against reports by philanthropists, social workers, policemen and others, which along with Booth's own assessments provided as scientific a record as was available at the time.[24] The survey was published in 1891 in four sheets as the *Descriptive Map of London Poverty 1889* (see Figure 3.5).[25]

Having completed the street survey, Booth also studied 4,000 households in detail to provide additional data on the causes of poverty. His conclusions were that poverty was multidimensional in its *aspect* (namely the observed data) and in its *causes*, and that there were causes that came from the state of the labour market – namely that they were circumstantial. The causes, both circumstantial and personal, fed off each other, so that external poverty might exacerbate personal circumstances and vice versa. Bearing this in mind, it is evident that despite the first impression of the map as being quite stark in dividing the city into only eight classes,

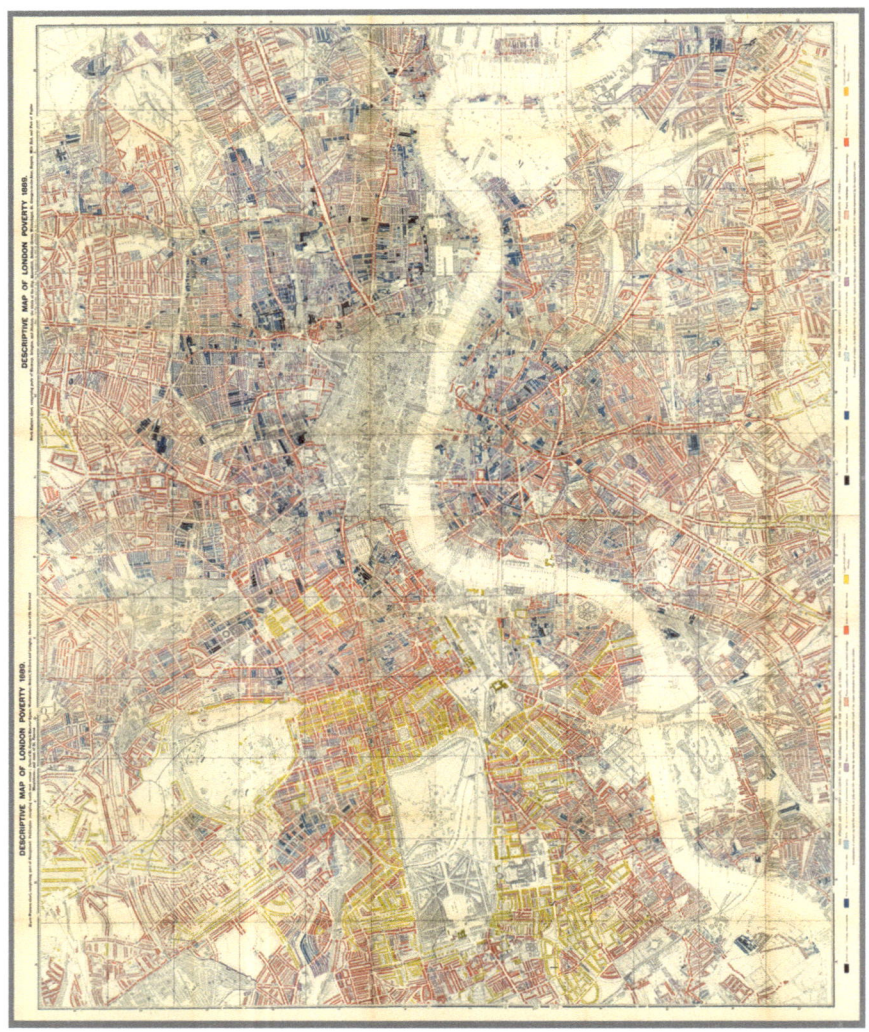

Figure 3.5 Charles Booth's *Descriptive Map of London Poverty*, 1889, sheets 1–4 compiled into a single image.

Drawn up to accompany C. Booth, *Labour and Life of the People*. Appendix to volume II, ed. Charles Booth (London and Edinburgh: William and Norgate, 1891).
Image copyright Cartography Associates, 2000.

in fact its scale of colours represented a combination of factors such as regularity of income, work status and industrial occupation. Booth was in effect recognising that in many instances regularity of income was primarily shaped by a person's occupation.

While scholars criticise the supposedly subjective nature of the survey (and its particularly moralistic language, in terming the lowest class as vicious – that is, pertaining to vice), the scope, rigour and scientific method make Booth in many scholars' eyes the first true social scientist.[26]

Booth's maps showed that pockets of poverty could be found throughout the city, while the East End working classes, which had been feared as a source of political action, was much more varied than had been assumed. In refutation of the public fears, it was clear that the poorer working classes were unlikely to be able to organise any form of action, let alone threaten social order. According to Kevin Bales, these results alone were seen as a breakthrough by many commentators,[27] while Booth's paper to the Royal Statistical Society in May 1887 led to widespread newspaper coverage due to his reported 'illumination of what had become in the public's mind as "darkest London"'.[28]

Ten years after the first survey Booth and his team undertook a revision of the maps: 'Every street, court and alley has been visited . . . changes have been most carefully considered . . . [most changes are] the result of the natural alterations of ten years of demolitions, rebuilding and expansion involving changes in the character or distribution of the population'.[29] The results of this survey were published in the maps 'Descriptive of London Poverty 1898–9', which comprised 12 sheets of detailed maps, covering a wider area still than the 1889 survey. In order to create this survey, members of the Booth Inquiry went on walks around the area, usually accompanied by a policeman, during the period between May 1897 and October 1900. The preceding decade had coincided with a period of quite extensive slum clearance across London and this shift in localised patterns of poverty and prosperity is very clear when comparing the two maps, as we will see below. Indeed, the notebooks recording these interviews are a vivid record of how the streets had changed since 1889.[30] See, for example, the page for an area around Chapel Market, Islington (Figure 3.6):

> Large business done, Jews have lately begun to take over the shops. Shopkeepers live above their shops & stalls. Two courts out of it on the North side – purple barred with black in the map. A roughish set of coster & fish curers – no trouble to the police. Dark rather light

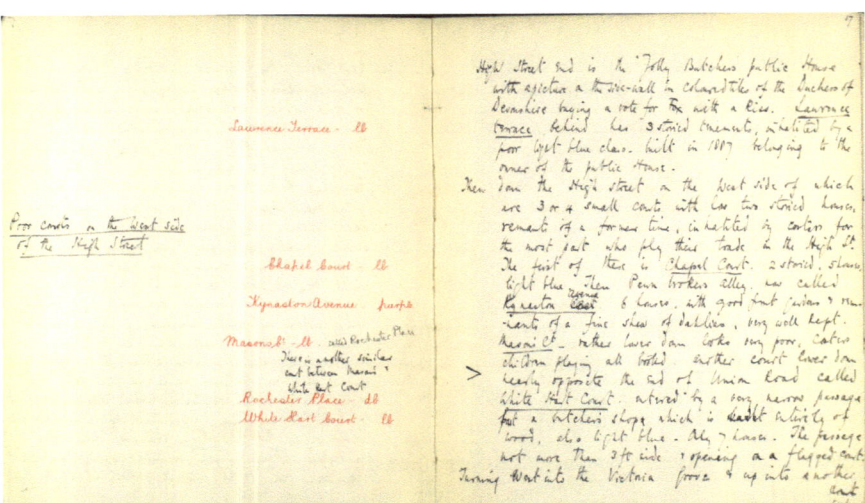

Figure 3.6 'George H. Duckworth's Notebook: Police and Publicans District 14 [West Hackney and South East Islington], District 15 [South West Islington], District 16 [Highbury, Stoke Newington, Stamford Hill]', 1897, p. 199.

Booth, C. 'Poverty Series Survey Notebooks (Online Archive).' British Library of Political and Economic Science, https://booth.lse.ac.uk/notebooks/, Reference: BOOTH/B/348.

to dark blue then black. Union Square has 15 2 storied cottages. Chapel Place is the other court.[31]

Reading the maps of poverty

The Inquiry used two sets of watercolour maps throughout the study to document the condition of each street in the study area and as a basis for recording initial findings before preparing the final version.[32] These were displayed to the public at two of the principal London Settlement Houses (Toynbee Hall and Oxford House),[33] with an invitation to the public to view them and propose corrections. The maps were also displayed to a scholarly audience; in his 1887 address to the Royal Statistical Society, Booth drew people's attention to the 'many coloured map hanging on the screen', encouraging them to draw close to see its detailed colouring, '[recommending] this graphic method of representing the condition of the people as one most easily to be comprehended, and, what is still more important, most easily to be verified'.[34] Booth was intending to use his graphic method to communicate the true nature and extent of poverty in London to both the public and the politicians of the time, to expose 'the numerical relations which poverty, misery and depravity bear to regular earnings and comparative comfort, and to describe the general conditions under which each class lives'.[35]

The survey and its maps drew extensive press coverage. *The Guardian*, for example, stated that Booth's map of poverty had lifted the 'curtain behind which East London had been hidden' and presented the nation with a 'physical chart of sorrow, suffering and crime'.[36] Booth's maps went beyond displaying graphic patterns; they provided the spatial and social context of poverty by showing the arrangement of one in relation to the other. He could illustrate how frequently poverty streets were situated cheek by jowl with red or gold streets. If we look at Figure 3.7, which zooms in on a patch of the 1889 survey maps, it is clear that even in the impoverished East End the well-to-do streets, coloured red, are just one turning away from the next step down in class – those coloured pink, who are just a street away from the blues and blacks of the bottom grades of street. This subtle organisation of the marginal separation of class or land use was carried out by integrating similar uses or classes along the same street alignment but effectively segregating different uses by putting them on different alignments.[37] Another typical feature of the London

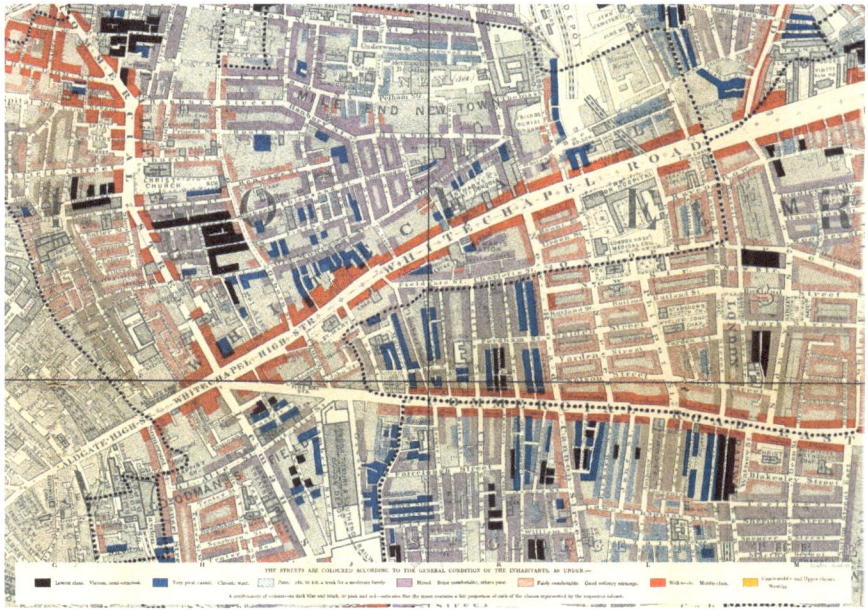

Figure 3.7 Sample area from *Descriptive Map of London Poverty*, 1889, showing marginal separation of poverty from relative prosperity (overlaid with the key to the map).

Booth, *Life and Labour of the People*. Image copyright Cartography Associates, 2000.

street is its domino-like symmetry of uses across each street, which, as Richard MacCormac has stated, 'affirms its character as a place'.[38] Yet, there was no explicit plan to organise the nineteenth-century city in this way. Instead, this pattern had emerged as part of the process of continuity and change which brought about the spatial logic of the city at the time.

The manner in which the historical city adapted to form marginal differences in accessibility that differentiate class and situation is striking. By using street morphology to organise the social and economic form of the city, a variety of classes could be located in the same area. These findings support the historical analysis which showed how the area's structure provided opportunities for a well-organised mesh of economic interdependency, so that diverse groups could benefit from their mutual spatial proximity to support a flexible spatial economy. Naturally, we should not forget that use of the nineteenth-century city's streets was demarcated by sex as well as class. Ellen Ross has written about how for women of Booth's poverty classes there were elaborate systems of borrowing, lending and support, even with small loans.[39]

David Reeder has described how much the 'territorial variations in the conditions of life in different London districts were related to the residential patterns that had evolved over the course of the century'. Considering the map in its entirety, it is striking how it is made up of distinctive patches of poorer streets, set within a frame of 'well-to-do' streets. Equally, London's then socio-spatial framework of an armature of prosperous streets (coloured red), demarcates the boundaries between one neighbourhood and the next. We can also see how at its edges the city does not fall off into an abyss of no-man's-land but has more porous street structures set within open land, challenging conventional descriptions of the urban fringe as *tabula rasa*.[40] The map was in effect capturing the process whereby London's social contours are shaped over time. But more significantly (as Kimball maintains), the visual rhetoric of the map changed the public view of poverty, making the problem seem much smaller than had been supposed, and thus manageable. We can see this type of rhetoric take effect in the comments of a Daily News reviewer:

> [The map] is in many colours, and the specks of black will show us where to find the haunts of the lowest class . . . Happily, the strange landscape shows a fair predominance of the more cheerful colours. It is a pink, and a red, and a light blue landscape, on the whole; and only here and there . . . are the dismal shades which seem but so many varieties of black.[41]

It is not only the fine-scale layout that seems to have had an impact on social conditions, creating pockets of irregularity in the urban grid. Pfautz notes how for Booth, '[the concept of] 'poverty area' represented one of the most developed and sophisticated areal concepts in the inquiry. Specifically, it denoted the little groups of 'black and blue' streets which, it will be remembered, had an apparently random pattern of distribution on the 'social map' of London.[42] Larger-scale obstacles in the urban fabric had a deleterious effect on the ability of people to move around and improve their social and economic conditions. Booth frequently noted in his writing that physical boundaries such as railways had the effect of isolating areas, walling off their inhabitants and isolating them from the life of the city. The urban historian H. J. Dyos has also pointed out 'how often these introspective places were seized by the "criminal classes", whose professional requirements were isolation, an entrance that could be watched and a back exit kept exclusively for the getaway . . .'.[43] Gareth Stedman Jones has also noted that

> [o]ne great effect of railway, canals and docks in cutting into human communities [is] a psychological one. . . East Londoners showed a tendency to become decivilised when their back streets were cut off from main roads by railway embankments. . . Savage communities in which drunken men and women fought daily in the streets were far harder to clear up, if walls or water surrounded the area on three sides, leaving only one entrance.[44]

Booth's maps show how the impact of railway lines was frequently, and ironically, to reinforce class division: on the one hand they created the possibility to escape the city to clearer air a commuting distance away for those who could afford the move, but on the other hand, for those who could not afford to move away, the situation was worsened by the impact of the railways on the local environment. Booth himself wrote of an area of East Battersea:

> Of the other extreme, the worse elements have for the most part taken refuge in blocks of houses isolated by blank walls or railway embankments, or untraversed by any thoroughfare. Some of the courts have long been notorious in the neighbourhood – one, for instance, is popularly known as 'Little Hell'.[45]

If we zoom in on the West End of London we find a clear example of the marginal separation of classes mentioned above, with a network of streets that ranges from the grandest houses to the blackest slums (Figure 3.8

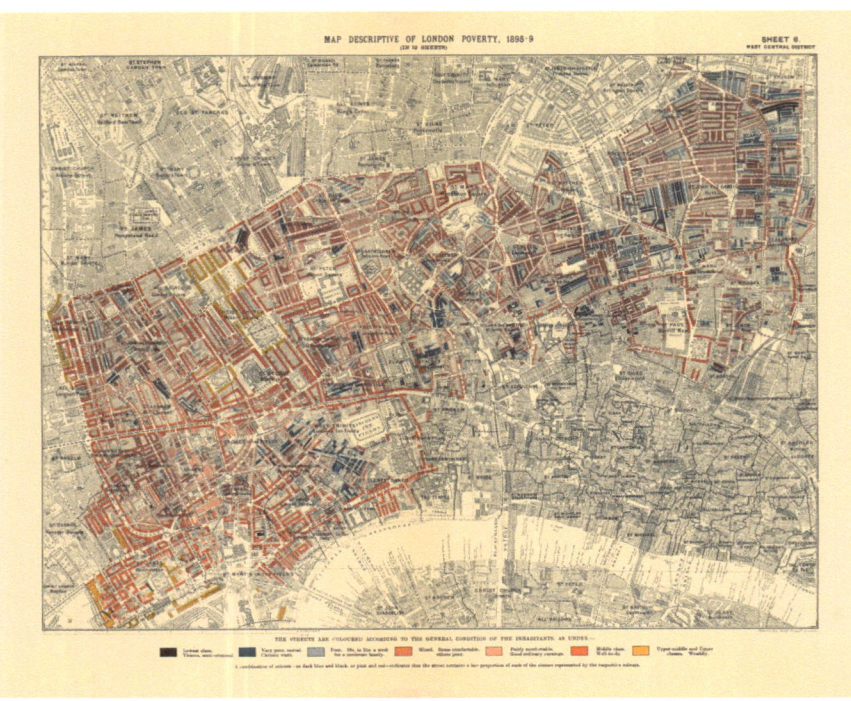

Figure 3.8 Charles Booth's *Map Descriptive of London Poverty*, 1898–9, sheet 6.
West Central District. Covering: Westminster, Soho, Holborn, Covent Garden, Bloomsbury, St Pancras, Clerkenwell, Finsbury, Hoxton and Haggerston.
LSE reference no. BOOTH/E/1/6.

and 3.9). The map also highlights how the parish of St Giles in the Fields to the south of Oxford Street (the street coloured red running west to east) had pockets of severe poverty, with a large number of streets in the poverty class colours of black, dark blue and light blue, framed by streets coloured red ('middle class'). As Christopher Breward has written, 'The stark difference between Marylebone, to the north of Oxford Street and Soho, to the south, was marked along architectural, ethnic, social and professional lines';[46] further, the poor Soho district was only a few turnings away from the streets coloured yellow (or gold) in the parish of St George Bloomsbury to the north of Oxford Street. Notably, the yellow streets tend to be removed slightly from the main streets of the city, by having only one flank of the square facing a well-connected street. At the same time (as already mentioned in Chapter 2 – see Figure 2.7), the specific morphology of Soho that made it an area which differs from its surroundings allowed it to contain a range of marginal social activities and classes which could coexist with the contrasting surrounding areas by virtue of the spatial containment of the district. This subtle

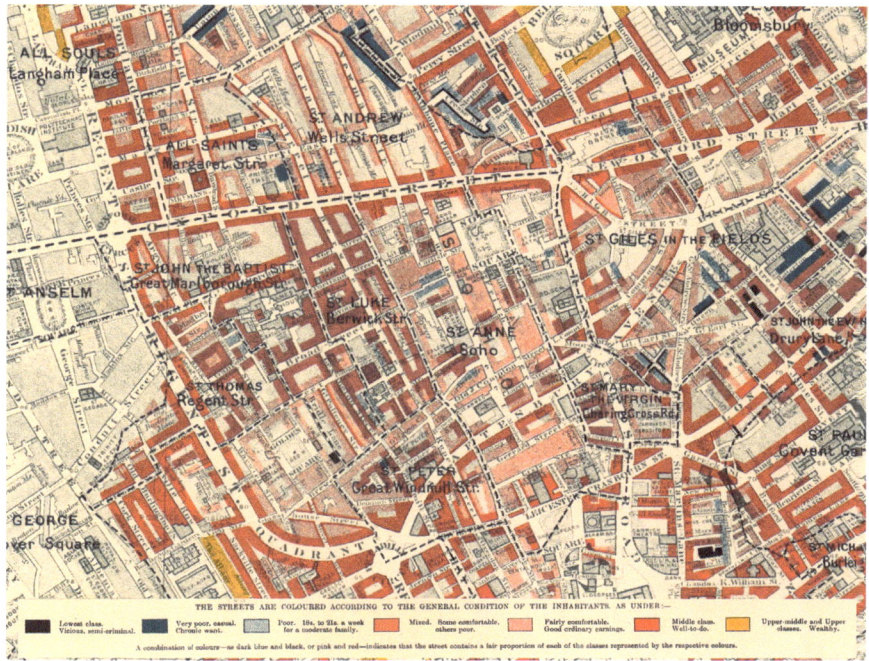

Figure 3.9 Detail of Charles Booth's *Map Descriptive of London Poverty*, 1898–9, sheet 6 (overlaid with the map's key).

LSE reference no. BOOTH/E/1/6.

organisation of space is the way in which the layout of streets has in the past intersected with the distribution of poverty in cities such as London, in a way quite different from modern-day gated housing areas.[47]

By translating the amorphous, ungraspable problem of poverty into a measurable social issue that could be targeted with pinpoint accuracy, Booth's maps had provided a means to start finding solutions to a problem that hitherto had been out of reach and out of sight. Booth, Dyos and Stedman Jones all pointed towards the possibility that changes to the urban layout could themselves contribute to the decline of poverty areas. The following section reviews the latest empirical research into this notion.

Analysing the maps of poverty

Booth's definition of poverty was intentionally relative, given that he was using a description of class, not income. In other words, he did not define a basic level of subsistence, below which an individual could fall; rather, he established what the conditions were in which poverty took place. Although he refers in his writings to a line of poverty, which he notionally positioned at a 'bare income' of 18s. to 21s. per week,[48] this was a hypothesised line demarcating the boundary between those who were just getting by and those who were in want. It was based on Booth's close observation of a statistical sample of a large population – across many industries active in London at the time. These conditions included the physical setting within which people were living. Pfautz also notes that one of the factors that was especially important for Booth 'in determining the class of residents . . . [was] *situation*, in contrast to *site*. Here Booth took specific note of the accessibility of areas to one another, particularly areas of living to areas of work.' In other words, Booth recognised that if an individual found it difficult to get to work from home, they would find it harder to get a job. Indeed, there are many accounts of dockers having to live within reach of the port of London, or the city's jobbing tailors having to live within barrow-wheeling distance from the tailoring industry's heartland.[49]

Booth's observations on poverty situated the problem as being to do with the regularity of income just as much as its level. Yet he also put great emphasis on the impact of street layout on social situation. Scattered through his writing are comments such as '. . . the "poverty areas" tended to be literally walled off from the rest of the city by barrier-like boundaries that isolated their inhabitants, minimizing their normal participation in the life of the city about them. . .'.[50] Booth was also deeply aware of the impact of the physical conditions of housing on poverty. Amongst his recommendations to the Royal Commission on Housing (1901) about the urban spatial solutions to housing and poverty were provision of better transport to allow for dispersal to the suburbs, improved planning – open space, widening of thoroughfares and opening up of courts, closing of houses not fit to live in, supervision of new buildings, slum clearance and a policy of construction and reconstruction throughout London (not only in its crowded parts).[51]

One of the worst areas was the infamous Nichol district in Shoreditch, East London. It had featured in a thinly fictionalised novel by Arthur Morrison (written at the instigation of a local vicar, Arthur Osborne Jay). Despite, or probably because of, its somewhat sensationalist language, *A Child of the Jago* did much to raise the public consciousness of the dire conditions in this location at the time:

> It was past the mid of a summer night in the Old Jago. The narrow street was all the blacker for the lurid sky; for there was a fire in a

farther part of Shoreditch, and the welkin was an infernal coppery glare. Below, the hot, heavy air lay a rank oppression, on the contorted forms of those who made for sleep on the pavement: and in it, and through it all, there rose from the foul earth and the grimed walls a close, mingled stink – the odour of the Jago.[52]

As Morrison wrote, the Nichol was a densely populated warren of streets containing appalling housing conditions. Life expectancy was said to be just 16. The district's rotten housing stock meant that it was the last refuge of the poor and its bewildering layout meant that it had a reputation locally as a criminal enclave. One resident attested that living in the Nichol was 'something like a ghetto . . . In the Nichol there seemed to be a wall enclosing you'.[53] When surveying it in the 1880s Booth's team had classified the area's streets at the lowest grades, black and dark blue (respectively connoting 'Very poor, casual, chronic want; lowest class' and 'Vicious [i.e. pertaining to vice], semi-criminal' respectively) – see Figure 3.10a, showing the area on the 1889 map.

Booth's project became part of the drive for reform which sought state intervention to relieve poverty conditions. The Nichol's housing was demolished and replaced by the Boundary Estate. It was the first project constructed by the newly formed London County Council, the first state social housing in the country. The process of spatial change, which would normally take a significant period to have an impact on social patterns of life, was much more rapid in cases such as this. The aim of the slum clearances was to tidy up the overly complex geometry of the street layout. Instead of the dense, labyrinthine layout of the Nichol a central circus ringed by red brick blocks of flats were constructed, with streets radiating out from the centre to connect with the surrounding area. Once constructed, the complex was rapidly inhabited. Less than 10 years later, the streets were classified by Booth's team as 'pink, fairly comfortable' (see Figure 3.10b and, for an image of one of the estate's streets, Figure 3.11). Even at its time it was a step up from the surrounding area. However, in the same unfortunate pattern seen today in many regeneration projects, the original inhabitants of the cleared streets had to move elsewhere as they could not afford to move into the new housing. This led to worsening conditions in the surrounding housing, which became more overcrowded.

Research into the spatial nature of this variation in levels of poverty has used space syntax methods for modelling and analysing space to quantify the geometric, topological and metric properties of the Booth maps of poverty to see if there are consistent relationships between spatial isolation (or 'segregation' in space syntax terminology) and levels of poverty.

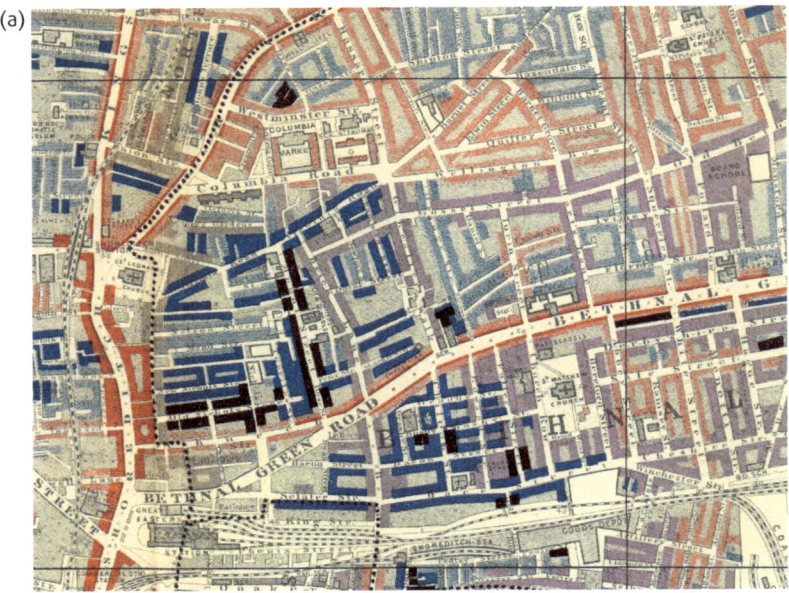

Figure 3.10 a) Detail of *Descriptive Map of London Poverty*, 1889 showing the Old Nichol area and b) Detail of the same area on the *Map Descriptive of London Poverty*, 1898–9, sheet 6, showing the new estate.

West Central District. Covering: Westminster, Soho, Holborn, Covent Garden, Bloomsbury, St Pancras, Clerkenwell, Finsbury, Hoxton and Haggerston.
LSE reference no. BOOTH/E/1/6.

Figure 3.11 Abingdon House, Boundary Estate, Old Nichol Street.

Photograph by Clem Rutter, Rochester, Kent (http://www.clemrutter.net) via Wikimedia Commons.

Space syntax methods use mathematical measures of a model of the street network to calculate the relative accessibility of every street to all other streets within the system (or within a defined distance). With these methods, measures of the relation of each street segment to all others can be set alongside social and economic measures, such as, in this instance, Booth's classifications.[54]

The first stages of space syntax research into the Charles Booth maps found that socially or economically marginalised individuals follow distinctive patterns of settlement and that underlying these patterns were spatial conditions that may have influenced this distribution. For example, the analysis found that interruptions to the grid structure significantly influenced the spatial configuration of a poverty area, giving rise to conditions of both spatial and social segregation.[55] Detailed spatial analysis found that while districts such as Soho had formed localised areas of poverty, the East End of London had a stronger differential between the spatial integration of 'middle class' streets and all the poorer streets. We can see this in Figure 3.12, which shows a section of the space syntax analysis of the Booth map of poverty 1889 for the East End, with the main streets classified as 'middle class' being markedly more likely to be coloured in the warmer shades of integration. In parallel, there were localised clusters of very poor streets, which the analysis found were physically cut off from the life of the city.

In showing a measurable relationship between spatial segregation and living in poverty, the research findings indicate that Booth's three poverty classes constituted a spatially defined poverty line. In other words, the space syntax analysis provided evidence to support Booth's own hypothesised line of poverty, with the streets coloured black, dark blue and light blue being much more likely than average to be spatially segregated.[56]

We saw in Chapter 2 how the spatial patterning of urban contagion was seen as akin to a diseased body. The diagnosis made through maps of disease was followed by the cure; the opening-up of the slum areas with wide thoroughfares which would overcome the social degradation that was perceived as being bound up in the environmental degradation of the city. Haussmann's redesign of Paris was a cleansing of disease, but also

Figure 3.12 Detail of *Descriptive Map of London Poverty*, 1889, showing the East End district of London, overlaid with space syntax analysis of spatial accessibility for each street to all other streets within 800m.

Image by the author.

a purging of pockets of revolutionary activity.[57] Similarly, John Nash's decision on the alignment of the new Regent Street in his masterplan from 1809–33 had been a conscious confirmation of the perceived need for separation between the 'streets and squares occupied by the nobility and gentry [to the west], and the narrow streets and meaner houses occupied by mechanics and the trading part of the community [to the east, broadly in the Soho district]'.[58]

Further analysis of change between Booth's maps of 1889 and 1899 showed that the slum clearance programme which had been implemented during that period had an effect of improving the physical and social situation of the immediate surroundings of the clearances, but this masked the fact that the poorest people contained within these areas had to find cheaper accommodation deeper within or outside of the district. In fact, detailed analysis of the spatial/economic change over time found that the areas surrounding the slum clearances experienced a marked drop in economic situation – a ripple effect as an outcome of spatial change and an indication that the improvement of spatial organisation may not have had a significant impact on the lowest classes.[59]

We have seen from the disease maps how spatial patterns of deprivation persist over considerable periods of time. Scott Orford and Danny Dorling's work is especially insightful in showing how the many attempts to improve housing quality over the past 100 years have 'failed to substantially alter the geography of poverty'.[60] Yet research into this persistence has tended to concentrate on the social causes of poverty, and much less on the possible effects of physical planning on poverty patterns. Other research has found distinctive patterns in the spatial distribution of poverty and that 'forms of deprivation are patterned spatially by a series of urban processes, which lead to greater concentrations of problems in particular places'[61] The notion that when poverty is concentrated in large areas, it can have deeper effects on its local population is important. In her study of these so-called neighbourhood effects, Ruth Lupton states that 'physical characteristics, through their impact on population mix, lead neighbourhoods to "acquire" certain other characteristics, such as services and facilities, reputation, social order and patterns of social interaction, as people and place interact'.[62] Areas can also acquire negative reputations that are very difficult to shake off. Problems such as access to resources, the opportunity to gain information on jobs or to improve one's education can be shaped by living in an area where the majority of the population lives in poverty. We saw this pattern in Hume's Liverpool, where he referred to the impact of the middle classes moving out, and indeed it continues today.

Space syntax research into the long-term persistence observed by Orford, Dorling and others selected one of the most highly deprived areas in London to see whether the persistence in poverty over extended periods of time is related to the spatial structure of the area. Government support in the form of housing benefit and council tax benefit were chosen as the most appropriate indicators of whether individuals were below a threshold of need, and hence the closest to Booth's own form of assessment. As contemporary data are an indication of situation below a notional poverty line, a different adaptation of data from the Booth dataset was needed. It was decided to create a proxy for Booth's method of classifying streets according to a scale in which the greater the proportion of benefit recipients in a street segment, the lower its classification. The poverty data were plotted against data on the space syntax measures of accessibility of the street network, to see if there was a relationship between the two. Whilst the study was small in its scope, its findings suggested that even today there is a correspondence between poverty (as measured by the proportion of households in a street being recipients of both benefits) and spatial segregation, although the wider trend of the area's gentrification means that the relationship is not as strong today as it was in the past. In fact, poverty is deepening in some pockets of the contemporary East End while prosperity is increasing elsewhere.[63]

The relative stability of concentrated disadvantage is a rather remarkable phenomenon. In 2006 *The Economist* provided intriguing examples of the persistent nature of deprivation in London from Booth's day to the present, even at a micro ecological level. It compared poverty and prosperity rates between 1898 and 2001 in a section of the Chelsea neighbourhood of west London. The study found that in general the classes had 'upgraded', especially along the southernmost edge of the neighbourhood and on the west. This was partly a reflection of the neighbourhood becoming popular in the 1960s with cultural figures. In some ways it is transformed, as the newspaper notes; today Booth's investigators would be recording the presence of designer clothing shops and a high concentration of expensive cars. Yet small pockets of poverty remain present. Where Booth's investigators observed 'Evil looking drink-sodden old . . . women' these 'have been replaced by the merely down and out or struggling. . .'.[64]

Booth's legacy

Charles Booth's enquiry established the importance of empirically derived evidence. He showed that districts such as the East End were not an undifferentiated morass of poor, criminal streets, but in fact contained a variety of classes, with finely differentiated deprivation situations. While his critics argue about the 'impressionistic' nature of his study, it was in fact his development of a methodology for social investigation which combined direct observation with statistics, drawing on both quantitative (statistical) and qualitative methods, that established the importance of using mixed methods for studying complex problems. Not only did he test poverty by various indicators such as income, overcrowding, educational attainment, servant-keeping and so on, he also incorporated unquantifiable social influences, such as church attendance.

The maps were a vital tool in this regard. In contrast with the sensationalist accounts of 'darkest London' and the well-meaning studies that had preceded Booth, the maps provided a method for visualising a problem – not only to allow for targeted solutions, but also to start to hypothesise on the underlying systemic causes for those problems and, thus, to make legislation that was built on evidence, not rhetoric. Booth also showed that data become more persuasive if mapped because they extend opportunities for interpretation beyond the domain of the statistician.

Booth's maps provided visual confirmation for his statistical evidence, showing the need for legislation to alleviate the situation of large cities such as London in general, as well as the East End in particular. In addition to his ground-breaking work on poverty in old age,[65] Booth also addressed the Royal Commission on Housing (1901) on the subject of urban spatial solutions to housing and poverty.[66] From the earliest days of his enquiry, Booth was concerned with both housing form and overcrowding, recognising the relationship between the physical environment and poverty:

> Space and air are everywhere at a premium . . . In the inner ring all available space is used for building, and almost every house is filled up with families. It is easy to trace the process. One can see what were the original buildings; in many cases they are still standing, and between them, on the large gardens of a past state of things, have been built the small cottage property of to-day. Houses of

three rooms, houses of two rooms, houses of one room – houses set back against a wall or back to back, fronting it may be on to a narrow footway, with posts at each end and a gutter down the middle. Small courts contrived to utilise some space in the rear, and approached by archway under the building which fronts the street. Of such sort are the poorest class of houses ... Another sort of filling up which is very common now is the building of workshops. These need no new approach, they go with, and belong to, the houses, and access to them is had through the houses. Some are even arranged floor by floor, communicating with the respective floors of the house in front by a system of bridges. These workshops may or may not involve more crowding in the sense of more residents to the acre, but they, in any case, occupy the ground, obstruct light, and shut out air.[67]

Several Building Acts followed, transforming the traditional London morphology. These Acts set out minimum permitted street widths and a maximum ratio of height to street width, banning courts, entrances closed off from the streets and dead-end streets. The rules led to building at higher densities, with greater distance between the blocks than before. Instead of building dense aggregations of two-storey houses arranged in courts and alleys, regulation determined that housing must be constructed with a setback from the road in front of the block to cope with the new height requirements, and with open space between the blocks at the rear. The new rules also prevented infill development due to the spacing restrictions. Height limits introduced further restrictions on building proximity (due to the need for air circulation). Regulations were also set governing the form of staircase and balcony access. Although balconies were highly valued by tenants, they were seen under the new regulations as an unhealthy mixing of people within a block. Finally, legislation was increasingly made about rooms and their layout – minimum sizes were set out and houses were ideally to be self-contained. For the first time, legislation explicitly defined the ways the buildings could be arranged and guaranteed that in the future there would no longer be rooms constructed without outside access, light and air. It also influenced planning thinking regarding the need for suburbanisation. Clearly new forms of housing alone were not going to solve the problem of overcrowding and, in his final volume of the survey, Booth argued for the city's suburban expansion as the one and only way that overcrowding, congestion, squalor, poverty, disease and degeneracy would be resolved.

This would take the form first of physical expansion and then the expansion of the city administratively.

Booth's conception of the city as being made up of natural areas with their own local ecology was highly influential. Although in the early part of his investigations Booth more or less confined himself to the use of pre-existing administrative areas, primarily for descriptive purposes, he later started to construct his own spatial units of analysis, using these to explain differences between different areas of a city. The natural area concept was a complex one, comprising a combination of several factors – site, situation, population type and institutions – that collectively determined its character. This concept was one of several reasons that Booth's project continued to have impact in the twentieth century, notably influencing the Chicago School.[68]

Subsequent social mapping projects – by the Hull-House settlement to record data on wages and nationalities in Chicago, and by W.E.B. Du Bois on the 'Negro Inhabitants' of a district of Philadelphia and their social condition and, back in London, by Booth's own team member, George Arkell, to record data on the relative density of settlement patterns in Jewish East London – are all direct legacies of Booth's work. All three studies will be elaborated in the next two chapters. More broadly, the legacy of Booth's work in social mapping can be seen even today across a wealth of topics, from disease through poverty, racial segregation and crime, as we will see onwards throughout this book.

NOTES

1. A. Morrison, *A Child of the Jago*, 3rd ed. (London: Methuen & Co, 1897).
2. An important exception to this is David Reeder's chapter in a book on nineteenth-century social investigation: D. Reeder, 'Representation of Metropolis: Descriptions of the Social Environment in *Life and Labour*,' in *Retrieved Riches: Social Investigation in Britain 1840–1914*, ed. D. Englander and R. O'Day (Aldershot: Ashgate, 2003). In fact, David Reeder's notes on the London Topographical Society's reproduction of Booth's first set of published maps are among the most detailed analyses of the maps. See D. Reeder, *Charles Booth's Descriptive Map of London Poverty, 1889* (London: London Topographical Society, 1984). See also the work of S. Swensen, 'Mapping Poverty in Agar Town: Economic Conditions Prior to the Development of St. Pancras Station in 1866,' (London: London School of Economics, 2006); and analysis of Booth's wealthy classes in a single district of London in P.J. Atkins, 'The Spatial Configuration of Class Solidarity in London's West End 1792–1939,' *Urban History* 17, no. -1 (1990).
3. D. Englander and R. O'Day, eds, *Retrieved Riches: Social Investigation in Britain 1840–1914 (Paperback Edition, 2003)* (Aldershot: Ashgate, 1998), p. 328.
4. W.S.F. Pickering, 'Abraham Hume (1814–1884). A Forgotten Pioneer in Religious Sociology,' *Archives de sociologie des religions* 17, no. 33 (1972), p. 35.
5. A. Hume, *Condition of Liverpool, Religious and Social, Etc*, 2nd ed. (Liverpool: Privately printed, 1858), p. 21.
6. Hume, *Condition of Liverpool*, p. 21.

7. The Act came about due to the findings of the Report on the Health of Towns (1840) that 39,000 persons in Liverpool lived in 7,800 underground cellars and 86,000 persons in the same city lived in 2,400 airless courts of back-to-back dwellings.
8. The map dates from 1845 and reveals the cramped conditions in the Broad Marsh area of Nottingham.
9. Hume, *Condition of Liverpool*, p. 41.
10. J. O'Brien and S. Griffiths, 'Relating Urban Morphologies to Movement Potentials over Time: A Diachronic Study with Space Syntax of Liverpool, UK,' in *11th International Space Syntax Symposium*, ed. Heitor, T. (Chair), M. Serra, J.P. Silva, A. Tomé, M.B. Carreira, L.C. Da Silva and E. Bazaraite, 98.1–98.11 (Lisbon, Portugal: University of Lisbon, 2017).
11. See O. Uduku and G. Ben-Tovim, *Social Infrastructure in Granby/Toxteth: A Contemporary Socio-Cultural and Historical Study of the Built Environment and Community in 'L8'* (Liverpool: Race and Social Policy Unit, University of Liverpool, 1998).
12. D. Smith, *Victorian Maps of the British Isles* (London: Batsford, 1985), pp. 63 and 110, although Rosemary O'Day attributes much of the technical innovation in social geography to George Arkell, who went on to draw up the map of Jewish East London, about which more in Chapter 5. R. O'Day and D. Englander, *Mr. Charles Booth's Inquiry: Life and Labour of the People in London Reconsidered* (London: Hambledon Press, 1993), p. 18. Along with the work of Anne Kershen and David Englander, Rosemary O'Day's book is one of the most comprehensive modern-day overviews of Booth's project.
13. T. Simey and M. Simey, *Charles Booth, Social Scientist* (London: Oxford University Press, 1960), p. 64.
14. Reeder, 'Charles Booth's Descriptive Map of London Poverty, 1889.'
15. J. Hanson, 'Urban Transformations: A History of Design Ideas,' *Urban Design International* 5 (2000), pp. 217–18.
16. A. Davin, *Growing up Poor: Home, School and Street in London 1870–1914* (London: Rivers Oram Press, 1996), pp. 34–5.
17. Rev. A. Mearns, *The Bitter Cry of Outcast London: An Inquiry into the Condition of the Abject Poor* (London: James Clarke & Co, 1883), p. 7.
18. Bales, 'Charles Booth's Survey,' p. 92.
19. C. Booth, 'The Inhabitants of Tower Hamlets (School Board Division), Their Condition and Occupations,' *Journal of the Royal Statistical Society* 50, no. 2 (1887).
20. Octavia Hill's meticulous work on housing conditions, highlighting for example the impact of 'middle men' in inflating the cost of renting housing, led the UK Government Commission in 1886 to look at the housing conditions of the working classes.
21. Beatrice Webb *née* Potter was a cousin of Booth's by marriage, a devoted writer on social matters. A member of the Fabian Society, she helped to establish the London School of Economics and Political Science with her husband, Sidney Webb, amongst others.
22. Although Hyndman's claim is said to have sparked Booth's enquiry, there is no evidence they met before 1886 and Booth's interest in poverty dates back much earlier.
23. The areas coloured purple might indicate a street in transition from one state to another, and indeed the space syntax comparison between the two maps has found that purple streets (at least within the East End study area) were more likely to change – whether up or down a class. Pfautz cites one example of this from Booth's police interview notebooks: 'Kensal New Town, owing to its distance from inner London, was not a popular district for artisans, but the opening up of the Central London Railway and the advent of electric trams have completely altered its outlook as a place of residence for the "pink" class.' H.W. Pfautz, ed. *On the City: Physical Pattern and Social Structure; Selected Writings of Charles Booth*, Heritage of Sociology (Chicago: University of Chicago Press, 1967), pp. 116–17.
24. Booth, *Labour and Life of the People*, vol. 1, p. 5.
25. The four sheets covered an area from Kensington in the west to Poplar in the east, and from Kentish Town in the north to Stockwell in the south – and published in subsequent volumes of the survey. These maps are collectively known as the *Descriptive Map of London Poverty 1889*. They use Stanford's Library Map of London and Suburbs at a scale of 6 inches to 1 mile (1:10560) as their base. Information from https://booth.lse.ac.uk/learn-more/what-were-the-poverty-maps. Accessed 2 August 2017.
26. For example, Topalov (1993) maintains that there is an interpretive quality to the Booth maps: first in the definition of class division, second in the possible subjective assigning of families to class categories. In addition to this, he criticises the fact that some of the data

were extrapolated from the individual (school records) to the family level. Nevertheless, the Visitors' information was 'cross-checked against those of philanthropists, social workers, policemen and others' (Englander and O'Day, p. 124).

27. K. Bales, 'Popular Reactions to Sociological Research: The Case of Charles Booth,' *Sociology* 33, no. 01 (1999).
28. Bales, 'Popular Reactions to Sociological Research', pp. 155–6. The first paper was Booth, 'The Inhabitants of Tower Hamlets.' Booth went on to become the Society's president.
29. C. Booth, *Life and Labour of the People of London*, vol. 1: East, Central and South London (London: Macmillan and Co, 1904), pp. 6–7.
30. *Maps Descriptive of London Poverty, 1898–9*. According to the LSE website (see note above), the 12 sheets – covering an area from Hammersmith in the west to Greenwich in the east, and from Hampstead in the north to Clapham in the south – were published in the survey volumes between 1902 and 1903. The maps use Stanford's Library Map of London and Suburbs at a scale of 6 inches to 1 mile (1:10560) as their base.
31. George H. Duckworth's Notebook: Police and Publicans District 14 [West Hackney and South-East Islington], District 15 [South West Islington], District 16 [Highbury, Stoke Newington, Stamford Hill], 1897. Booth, C. 'Poverty Series Survey Notebooks (Online Archive).' British Library of Political and Economic Science, https://booth.lse.ac.uk/notebooks/. Reference: BOOTH/B/348, p. 17.
32. Kimball, 'London through Rose-Colored Graphics,' p. 366. Kimball writes further of how 'the first set, now held at the Museum of London, was watercoloured between 1886 and 1891; these were the foundation of the 1889 East End and 1891 London lithographic maps. The second set, now held at the London School of Economics, was watercoloured between 1894 and 1899, forming the basis of the final 1902 lithographic maps'.
33. Toynbee Hall was the first of many settlement houses, principally in the UK and the USA. These were social settlements set up by a movement whose purpose was to have middle class people live amongst the poor, sharing knowledge and skills and providing services, such as day-care, health and education.
34. Booth, 'Condition and Occupations of the People of East London and Hackney', pp. 284–5.
35. Charles Booth, 1889, quoted in Reeder, 'Charles Booth's Descriptive Map of London Poverty, 1889.'
36. *Manchester Guardian*, 17 April 1889, cited in A. Kershen, 'Henry Mayhew and Charles Booth: Men of Their Time,' in *Outsiders & Outcasts: Essays in Honour of William J. Fishman*, ed. G Alderman and C. Holmes (London: Duckworth, 1993), p. 113.
37. See more on this conceptualisation of marginal separation in B Hillier, 'Cities as Movement Economies,' *Urban Design International* 1, no. 1 (1996).
38. R. MacCormac, 'An Anatomy of London,' *Built Environment* 22, no. 4 (1996), p. 308.
39. E. Ross, 'Survival Networks: Women's Neighbourhood Sharing in London before World War I,' *History Workshop*, no. 15 (1983).
40. L. Vaughan, S. Griffiths and M. Haklay, 'Chapter 1: The Suburb and the City,' in *Suburban Urbanities: Suburbs and the Life of the High Street*, ed. L. Vaughan (London: UCL Press, 2015).
41. *East London Life Daily News*, 16 April 1889, Charles Booth Archive, London School of Economics, A58.53. Cited in Kimball, note 25.
42. Dyos, 'The Slums of Victorian London,' p. 25.
43. D. Cannadine and D. Reeder, eds, *Exploring the Urban Past: Essays in Urban History by H. J. Dyos* (Cambridge: Cambridge University Press, 1982).
44. G. Stedman Jones, *Outcast London: A Study in the Relationship Between Classes in Victorian Society* (Oxford: Peregrine Penguin Edition, 1984), pp. 15–16.
45. Booth, *Life and Labour of the People of London*, 1: East, Central and South London.
46. C. Breward, 'Fashion's Front and Back: "Rag Trade" Cultures and Cultures of Consumption in Post-War London C. 1945–1970,' *The London Journal* 31, no. 1 (2006), p. 30 – in a section entitled 'Carnaby Street schmutter'.
47. See detailed analysis of poverty and prosperity in the Oxford Street area of the 1880s and 90s in Vaughan, 'The Spatial Form of Poverty in Charles Booth's London'.
48. The quote is from Booth's address to the Royal Statistical Society, 1887, cited in A. Gillie, 'The Origin of the Poverty Line,' *Economic History Review* 49, no. 4 (1996, November), p. 715. See also P. Spicker, 'Charles Booth: The Examination of Poverty,' *Social Policy & Administration* 24, no. 1 (1990).

49. Anne Kershen, personal communication, 4 December 2017. Kershen reports that the 1930s/50s trade unionist Mick Mindel stated similar. It relates to the fact that around the turn of the twentieth century garments were wheeled in barrows or even prams to and from tailoring workshops and the premises of wholesalers.
50. Quoted in Pfautz, *On the City*, 120.
51. C. Booth, *Improved Means of Locomotion as a First Step Towards the Cure of the Housing Difficulties of London* (London: Macmillan, 1901).
52. Morrison, *A Child of the Jago*, p. 1.
53. R. Samuel, ed. *East End Underworld: Chapters in the Life of Arthur Harding* (London: Routledge & Kegan Paul, 1981), p. 2.
54. See appendix for further detail on the space syntax analysis of the Booth maps.
55. L. Vaughan et al., 'Space and Exclusion: Does Urban Morphology Play a Part in Social Deprivation?,' *Area* 37, no. 4 (2005).
56. Vaughan and Geddes, 'Urban Form and Deprivation.'
57. 'The effectiveness of riot or insurrection depends on three aspects of urban structure: how easily the poor can be mobilized, how vulnerable the centres of authority are to them, and how easily they may be suppressed. These are determined partly by sociological, partly by urbanistic, partly by technological factors.' E. Hobsbawm, 'Cities and Insurrections,' *Ekistics* 27, no. 162 (1969), p. 304.
58. J. White, *Some Account of the Proposed Improvements of the Western Part of London: By the Formation of the Regent's Park, the New Street, the New Sewer, &C. &C. Illustrated by Plans, and Accompanied by Critical Observations*, ed. J.E. Moxon (London: W. & P. Reynolds, 1814), p. 48. Dyos, 'The Slums of Victorian London,' p. 36 details how the poorer streets were 'deliberately skirted' by Nash's plans. A contemporaneous report on progress with the road works stated that the 'New Street' would create an opportunity for the Crown to preserve 'that best built-part of the Town from the annoyance and disgrace which threaten it on either side'. See Great Britain. Parliament. House of Commons, 'The First Report of the Commissioners of His Majesty's Woods, Forests, and Land Revenues [Electronic Resource]: In Obedience to the Acts of 34 George III. Cap.75. and 50 George III. Cap. 65,' (London: Author, 1812), Part 1, p. 96.
59. For detail of the analysis, see L. Vaughan, 'The Relationship between Physical Segregation and Social Marginalisation in the Urban Environment,' *World Architecture* 185, special issue on space syntax (2005).
60. Orford et al., 'Life and Death of the People of London,' p. 34.
61. P. Spicker, 'Poor Areas and the "Ecological Fallacy",' *Radical Statistics* 76 (2001), p. 3.
62. R. Lupton, '"Neighbourhood Effects": Can We Measure Them and Does It Matter?,' (London: London School of Economics, 2003), p. 5.
63. Vaughan and Geddes, 'Urban Form and Deprivation'.
64. *The Economist*, 'There Goes the Neighbourhood,' 4 May 2006.
65. See C. Booth, 'Poor Law Statistics,' *The Economic Journal* 6, no. 21 (1896); C. Booth, *Old Age Pensions and the Aged Poor: A Proposal* (London: Macmillan and Company, 1899); D. Collins, 'The Introduction of Old Age Pensions in Great Britain,' *The Historical Journal* 8, no. 2 (1965). Booth's concern about the relationship between old age and poverty led him to push for an old age pension in the form of a universal scheme of governmental support in old age, but there was concern that the 'undeserving poor' would waste all the money on drink. The legislation that followed was a somewhat watered-down version, but still, for its time it was an enlightened development.
66. Booth, *Improved Means of Locomotion*.
67. Booth, 'Condition and Occupations of the People of East London and Hackney', 281–2.
68. The 'Chicago School' connotes a group of sociologists who practiced at the University of Chicago in the first half of the twentieth century, whose work was distinctive in its formal systematic approach to gathering and analysing social data, an empirical approach that differed from the prevailing philosophical approach. See more about the School in Chapter 6.

4
Poverty mapping after Charles Booth

> While the buildings housed more people than before, it was much healthier. But the dwellings were inhabited by the better-class workmen and artisans. The slum people had simply drifted on to crowd other slums or to form new slums.[1]

Rowntree's study of town life in York

The previous chapter closed with an overview of Charles Booth's impact on legislation regarding poverty and housing in the years following his inquiry. Although his study had been widely acclaimed, it had its critics. The Liberal Party politician, Charles Masterman, was one of these. He had spent time himself living in the London slums, and wrote an impressionistic account of his time there in articles published collectively as *From the Abyss*. Rosemary O'Day quotes his sniffy description of Booth's 'maps of picturesque bewilderment of colour, infinite detail of streets and houses and family lives. And at the end of it all the general impression left was of something monstrous, grotesque, inane, something beyond the power of individual synthesis: a chaos resisting all attempts to reduce it to orderly law.' Masterman also compared Booth's work unfavourably with Benjamin Seebohm Rowntree's York study, which Masterman found to be better connected and 'more helpful'.[2]

Putting aside how reasonable it is to diminish the importance of Booth's maps, let alone the majesty and scope of his published study, the longer view of history normally brackets Booth and Rowntree together, given their collective influence on changing attitudes towards poverty as well as their impact on new legislation in this area.

Rowntree's York enquiry was conducted with a team of assistants during 1899–1900 and then published in 1901 under the title *Poverty, a Study of Town Life*. His study was inspired by Booth, but whereas Booth's work was extensive and used multiple sources, Rowntree chose to conduct an intensive study of a single town, much smaller than Booth's London.[3]

Despite the differences in scope, the York study found very similar results to those of Booth: large families crammed into small rooms without sanitation or ventilation, and disease rife in poverty areas. Rowntree's team found that a quarter of all the children living in York's slums died before the age of one and that, overall, the poverty rate of the town was at least as bad as that found by Booth in London. Even if a child survived poverty in childhood, they were likely to remain poor, whether due to the casualisation of labour or to sickness or injury from work. If one survived working life, old age was likely again to return a person to poverty; just as Booth had found, old age was closely correlated with poverty. Unsurprisingly, both Booth and Rowntree became campaigners for old age pensions.

Rowntree's report is full of tables of statistics regarding factors such as household income and the cost of the diet and cooking fuel required to feed a manual labourer (he contrasts manual labourers with 'brain workers'). By aiming to determine the nature of living in poverty, Rowntree was in effect investigating whether it was due to 'wasteful expenditure', or 'insufficient means'. Rowntree's report did much to reiterate Booth's findings that poverty was not, as was typically thought, a matter of fault; critically, his determination of a poverty line based on income was set at the point at which income was only enough for the 'maintenance of merely physical efficiency', as opposed to access to enough to allow for 'expenditure necessary for the development of the mental, moral and social sides of nature'.[4] Rowntree showed that poverty distress was caused principally by low wages or irregular work. Nevertheless, his report's map of licensed houses (which he refers to in the text as the 'drink map', see Figure 4.1) emphasised the problematic relationship between drink and poverty, which we will see more of in Chapter 6, where a lack of sobriety was commonly associated with crime as well as poverty. Here, the association is more nuanced. While Rowntree was critical about the poorest of the poor wasting their money on drink, he was also aware that consumption of alcohol in York was no greater than elsewhere in the country.[5]

Rowntree strove to emphasise how precariously the poor sat on a finely balanced point between just getting by and destitution. His compassion

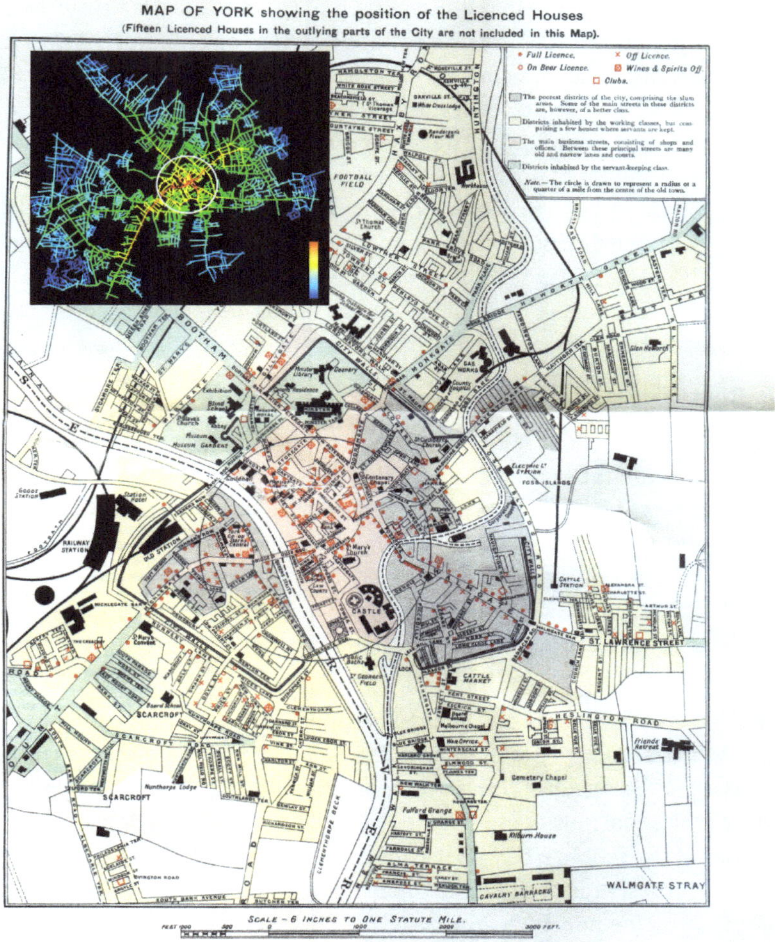

Figure 4.1 *Map of York Showing the Position of the Licensed Houses*, 1901. The map is shaded in four colours denoting the principal classes from lilac 'The poorest districts' to green 'Districts inhabited by the servant-keeping class'.

Inset showing space syntax map of city-wide accessibility in York.
Benjamin Seebohm Rowntree, 1901. Image of map of York courtesy Chris Mullen; space syntax map by Kayvan Karimi.

shines through his descriptions, for example, of what 'merely physical efficiency' constitutes in reality:

> And let us clearly understand what "merely physical efficiency" means. A family living upon the scale allowed for in this estimate must never spend a penny on railway fare or omnibus. They must never go into the country unless they walk. They must never purchase a halfpenny newspaper or spend a penny to buy a ticket for a popular concert. They must write no letters to absent children, for they cannot afford to pay the postage. They must never contribute anything to their church or chapel, or give any help to a neighbour which costs them money. They cannot save, nor can they join sick club or Trade Union, because they cannot pay the necessary subscriptions. The children must have no pocket money for dolls, marbles, or sweets. The father must smoke no tobacco, and must drink no beer. The mother must never buy any pretty clothes for herself or for her children, the character of the family wardrobe as for the family diet being governed by the regulation, "Nothing must be bought but that which is absolutely necessary for the maintenance of physical health, and what is bought must be of the plainest and most economical description." Should a child fall ill, it must be attended by the parish doctor; should it die, it must be buried by the parish. Finally, the wage-earner must never be absent from his work for a single day.[6]

Rowntree's comments on the drink map are incisive. He notes that the highest concentration of pubs is in the oldest section of the town, within and around the walls. This may, he surmised, be due to the town having served as a coaching centre, but the historical explanation did not solve in his mind the problem of there being an excess of drinking establishments in the poorest area of the centre. In fact, he devotes considerable space to analysis of the number of drinking establishments per population in his 'public houses' section of the book's supplementary chapter. The character of many of the poverty area's pubs, as being exclusively for drinking, are, he states, one of the causes of the prevalence of 'vertical drinkers', who are more likely to be heavy drinkers. The lower density of pubs outside of the centre, he argues, is mostly to do with the reluctance of magistrates to grant licenses to new establishments. Lastly, the change in the practice of organisations such as Trade Unions and Friendly Societies to meet in coffee houses instead of pubs (meaning a reduction in the use of pubs as community meeting places, which was a common feature earlier in the nineteenth century) had led to a narrowing of activities within the public house, although music and games, he reported, remained commonplace.

The spatial distribution of pubs in the poorer, central district is in fact a typical pattern in slum areas of the country at the time.[7] Several factors would have been at play, such as the pub providing warm, dry premises when the home was anything but, and the pub serving as a place for socialising outside of the home. It is evidently not a coincidence that pubs proliferated more on poor streets, while York's 'best' central streets, Monkgate, Clifton and Bootham, had hardly any.

Interestingly, space syntax analysis of York by Kayvan Karimi confirms the historical descriptions of these latter streets as being inhabited by the city's prosperous classes (which in the nineteenth century would typically have lived on the main avenues of towns and cities). Karimi's study investigated the character of six naturally evolved historic cities – York, Bristol, Norwich, Canterbury, Hereford and Winchester – in the few decades before Rowntree's study.[8] It found there to be a spatial structure common to all the cases examined; all had a compact network of streets at the centre, forming a core of connected streets which were the meeting point of linear routes from the outside of the town (see Figure 4.1 inset, which colours city-wide accessibility from red to blue, integrated to segregated). This core, marked in the white circle in the illustration, was always highly integrated spatially (namely, most accessible from everywhere to everywhere else and more likely to be busy with pedestrian and vehicular activity). The central cores of the six towns contained the most important urban functions, such as the market or the cathedral.

Karimi also found that the main roads into the town were highly accessible to principal routes within the city. The main roads into town were also the ones with the fewest pubs on Rowntree's map. Looking at their spatial structure, they only turned into local streets – typically lined with shops and other businesses – as they narrowed down at the town's heart; transforming themselves from routes *into* the town to routes *within* the town. The Clifton/Bootham alignment, for example, continued straight into High Petergate and, with a small kink that would have slowed traffic down, into Low Petergate, at which point the higher density of pubs was scattered amongst a much tighter mesh of shorter streets constituting the 'live centre' of the town, its commercial heartland (see Figure 4.2). In contrast, Walmgate, running into York from the south-east, had a much lower space syntax integration value and, unlike the other main arteries, it had a large number of pubs on its last stretch into town.

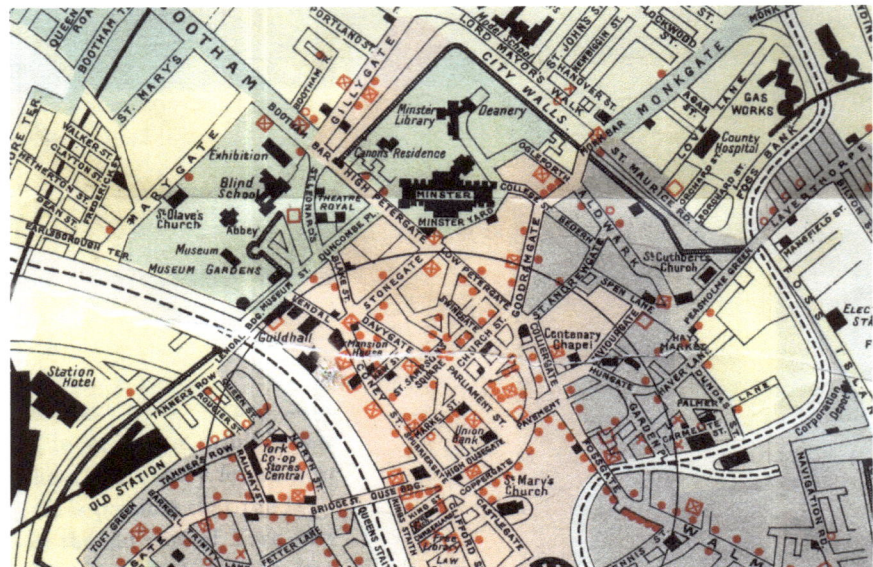

Figure 4.2 Detail of *Map of York Showing the Position of the Licensed Houses*, 1901. Rowntree, 1901. Image courtesy Chris Mullen.

This analysis tells us something interesting about how the nineteenth century city was formed: to shape and to be shaped by the pattern of social and economic activities within it. Clearly, the two were intertwined. The working class urban street in this period was much more active than in the past, resulting in the street becoming a place for informal collective life. As Brian Harrison has stated in his classic chapter on Victorian pubs,

> all but the busiest streets at that time united rather than divided the community: in working-class areas the emphasis is not so much on the individual home, prized as this is, as on the informal collective life outside it in the extended family, the street, the pub and the open-air market . . . the pavements were alive with pedestrians many of whom felt obliged to subscribe to drinking customs on the way. Many people earned their living on the pavement – the beggars, stallholders, acrobats, organ-grinders, pedlars.[9]

Harrison explains how the pub had a special role to play in the public sphere, in that it offered a place whose interior was masked by its frosted glass, a refuge from the street. It is no wonder that the temperance reformers, battling against the nineteenth-century drink culture struggled to compete with this enticing prospect, as we will see in Chapter 5.

Rowntree's team made close observations of activities in and around York's pubs. In one, he notes how the publican provides entertainment in a typically 'brilliantly lit, and often gaudily' decorated room, kept 'temptingly warm' in winter:

> At intervals one of the company is called on for a song, and if there is a chorus, everyone who can will join in it. Many of the songs are characterised by maudlin sentimentality; others again are unreservedly vulgar. Throughout the whole assembly there is an air of jollity and an absence of irksome restraint which must prove very attractive after a day's confinement in factory or shop.

Yet Rowntree's report shows him to have been well aware of the harm of excess drink, not only to the working man's pocket and his family's bellies. The way in which young children were exposed to the seedier aspects of life by being sent to fetch jugs of beer, let alone the use of pubs for prostitution, were also part of his first-hand accounts of life in York.[10]

Rowntree's report was celebrated as one of the most important empirical social studies of its time. By confirming Booth's findings as being true just as much for a small town as for a large city, he helped steer national legislation towards improving the lot of the poor. His definition of a poverty line, along with statistically founded calculations of the cost of basic nutrition needs, were taken up by the burgeoning Liberal Party, ultimately leading to legislation (picking up where Booth left off) on old age pensions and national insurance.

Hull-House and the wage map of Chicago, 1899

Meanwhile in the United States, following a visit by the social reformer Jane Addams to Toynbee Hall in London, a settlement house called Hull-House was founded in Chicago in 1889 by Addams and her colleagues, including Ellen Gates Starr. The house was located on the western edge of a densely settled multi-ethnic immigrant neighbourhood, at the intersection of Halstead and Polk Streets (immediately to the west of the map in Figure 4.3). It aimed to provide a variety of social services, education and guidance, to local families.[11] Its philanthropic activities also comprised research, including a nearly unique study into the daily lives and living patterns of Hull-House's neighbours.

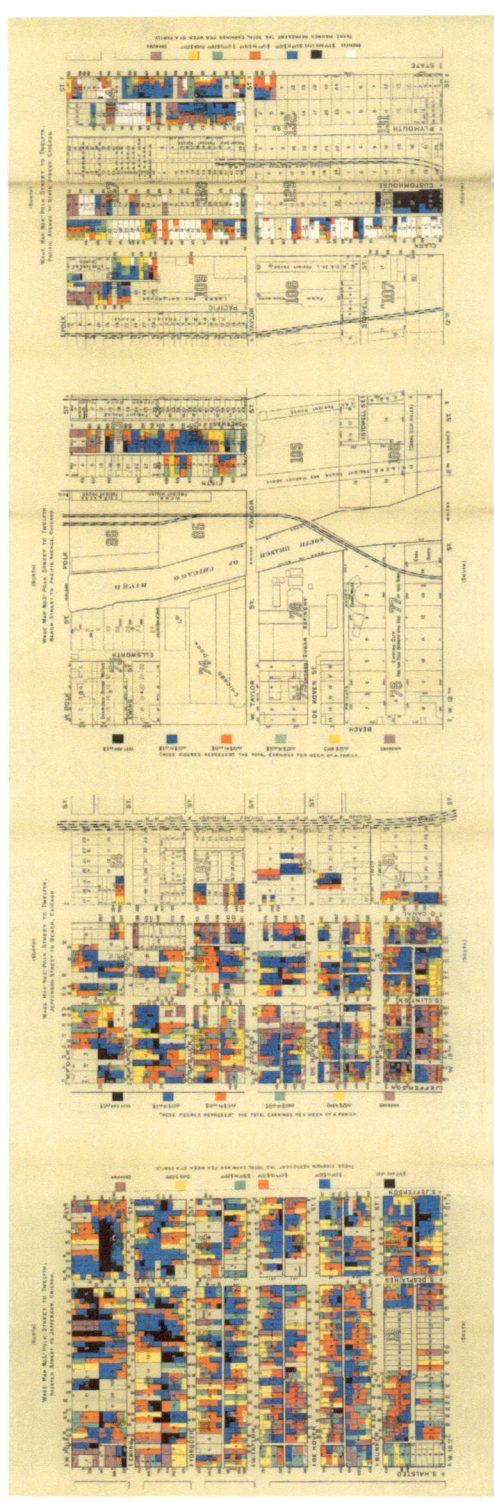

Figure 4.3 Hull-House wages maps 1–4, 1895.

Copyright Cornell University – PJ Mode Collection of Persuasive Cartography.

This was a period during which Chicago had grown at speed, with a much more dispersed pattern than we saw in New York in Chapter 2. There was little purpose-built housing, partly because it was expected that it would be wasteful to construct housing when factories and railroad terminals would soon take their place. Instead of Jacob Riis' New York setting of six- or seven-storey walk-up apartment blocks, the typical housing form in Chicago's poorest districts was made up of shoddily built dwellings, creating highly unsanitary housing that became increasingly overcrowded with the mass influx of immigrants into the city from the 1880s onwards. Districts such as the one studied here would typically have subdivided dwellings as well as stand-alone 'rear houses' constructed cheaply on the back of lots, accessible only through narrow passageways, forming totally inadequate, poorly lit and ventilated housing:

> Although poor buildings bring in such high rents that there is no business profit in destroying them to build new ones, the character of many of the houses is such that they literally rot away and fall apart while occupied. New brick tenement houses constantly going up replace wooden ramshackle ones fallen into an uninhabitable state. The long, low house on the northeast corner of Taylor and Jefferson cannot last long. No. 305 Ewing is in a desperate condition, and No. 958 Polk is disintegrating day by day and has been abandoned...
>
> Where temporary shanties of one or two stories are replaced by substantial blocks of three or four, the gain in solidity is too often accompanied by a loss in air and light which makes the very permanence of the houses an evil. The advantages of indifferent plumbing over none at all, and of the temporary cleanliness of new buildings over old, seem doubtful compensation for the increased crowding, the more stifling atmosphere, and the denser darkness in the later tenements.[12]

Within a few years of its founding, the residents of the settlement collectively produced the Hull-House Maps and Papers, which comprised a set of essays alongside a spectacular pair of maps (each a series of four sheets; see Figures 4.3 to 4.5 in this chapter and Figures 5.6 and 5.7 in Chapter 5), one on wages and the other on nationalities, which constitute an important visual record of the district at the time.[13] We will see more of the nationalities maps in the next chapter, but it is worth noting that they capture data in the same way as the wages maps, using the building lot as the unit of analysis, rather than the street or the street segment as in the case of Booth's maps.

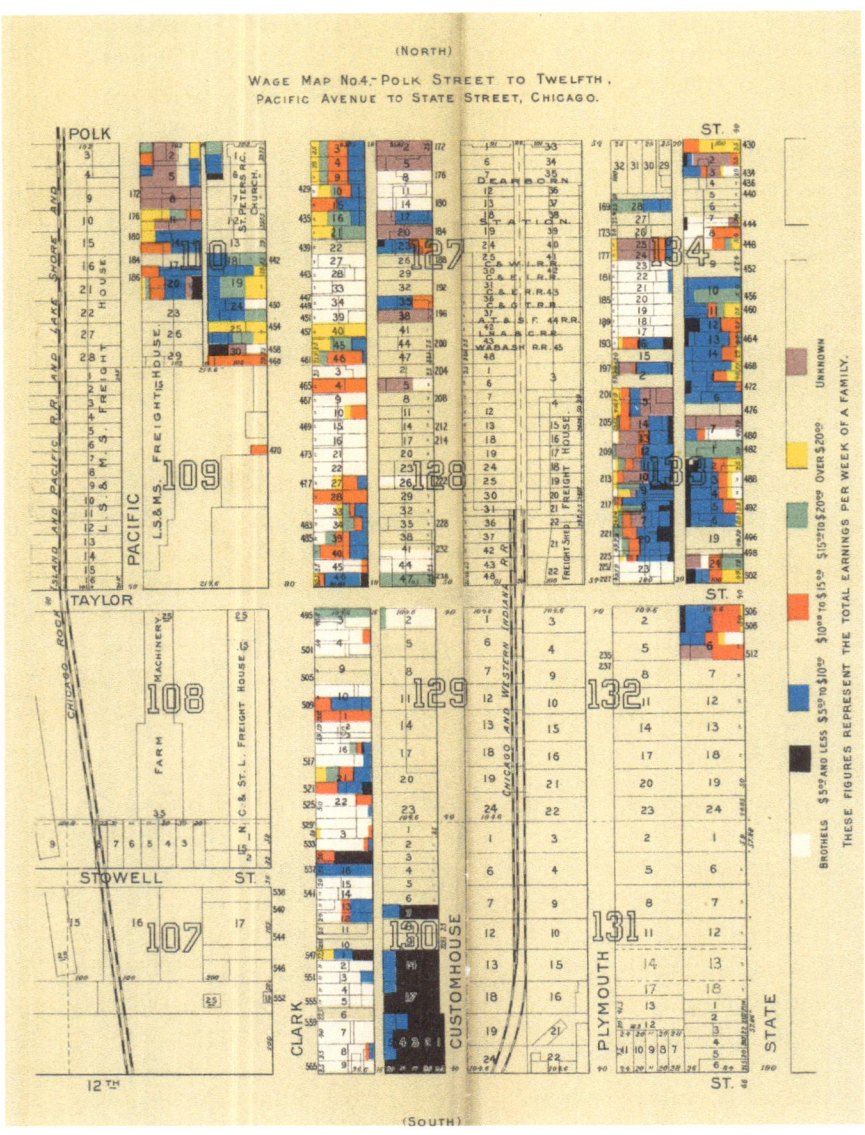

Figure 4.4 Detail of wage map no. 4, Hull-House wages maps, 1895.
Copyright Cornell University - PJ Mode Collection of Persuasive Cartography.

The study area covered Halsted Street on the west, State on the east, Polk on the north and Twelfth on the south, which is essentially the district running immediately east of the address of Hull-House itself. While the streets closest to Hull-House were 'the poorest and probably the most crowded section of Chicago', the study area also included 'east of the river a criminal district which ranks as one of the most openly and flagrantly vicious in the civilized world'.[14]

The data for the maps were compiled by agents working for Florence Kelley in the spring of 1893, visiting every house, tenement, and room in the Nineteenth Ward, collecting data about tenement inhabitants by interviewing them personally. Kelley conducted the investigation on behalf of the federal government, a *Special Investigation of the Slums of Great Cities*, living throughout its duration at Hull-House. Part of the reasoning behind the study was to capture an area in its transitional state, before it descended into the squalor seen in older cities in the US:

> In such a transitional stage as the present, there is surely great reason to suppose that Chicago will take warning from the experience of older cities whose crowded quarters have become a menace to the public health and security. The possibility of helping toward an improvement in the sanitation of the neighbourhood, and toward an introduction of some degree of comfort, has given purpose and confidence to this undertaking. It is also hoped that the setting forth of some of the conditions shown in the maps and papers may be of value, not only to the people of Chicago who desire correct and accurate information concerning the foreign and populous parts of the town, but to the constantly increasing body of sociological students more widely scattered.[15]

The area of study was chosen as it was deemed to be the most representative of 'slum . . . conditions of life' in the city.[16] The data were then transferred onto the maps, capturing in graphic form the physical dimension of the distribution of weekly wages as well as the range and distribution of ethnic groups. The legacy of Charles Booth is obvious in the use of colour coded maps, instead of relying on narrative reports or statistical tables, to provide an omniscient perspective of the perceived problem of the urban slum. This allowed for visual analysis of the socio-spatial dynamics of the city at the time. Notably, the Hull-House researchers followed his method in cross-checking the house-to-house survey with statements by workers in the various trades and occupations,

> although the eyes of the world do not centre upon this third of a square mile in the heart of Chicago as upon East London when looking for the very essence of misery, and although the ground examined here is very circumscribed compared with the vast area covered by Mr. Booth's incomparable studies, the two works have much in common.[17]

The map's colour code indicated the weekly wages of the residents in one of five separate classes, from under $5 per week to over $20 per week (a category 'largely composed of land and property owners, saloon and shop keepers, and those in business for themselves'), although information on income would not necessarily have captured irregularity of employment. As well as the five wages classes, the map also recorded the location of brothels. Their location east of the river is quite an unusual feature, given that it points to an area physically separated from the immigrant quarter, with some of the poorest housing (Figure 4.4).

The level of detail on the maps, as well as their size (the four sheets together were 36 x 112cm), would have made an imposing sight.[18] The way in which the maps distinguished between more prosperous housing on the front of lots and the crowded 'rear houses' at the back allowed the observer to get to grips with the scale of the problem of housing as well as its spatial juxtaposition.

Even without undertaking detailed space syntax analysis, a study of the spatial distribution of poverty in just one of the sheets, *Wage Map No. 2 – Polk Street to Twelfth, Jefferson Street to Beach, Chicago*, is illuminating. By calculating the proportion of each class present on all the street alignments, it is possible to compare the distribution of the poorer classes on the main streets as opposed to the alleys (see detail of map in Figure 4.5). The results show that while the lowest two wage classes, black and blue, are present in less than 11 per cent of all street-facing buildings (0.76 and 10.13, respectively), they constitute nearly 71 per cent (21.7 and 49.06 per cent, respectively) of all buildings

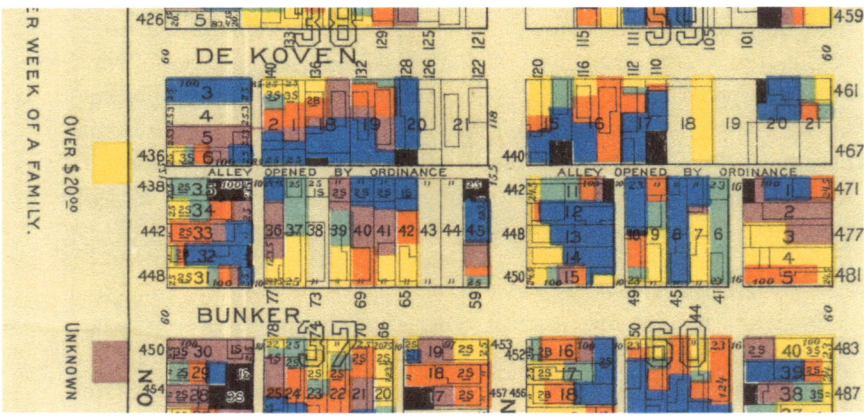

Figure 4.5 Detail of wage map no. 2, Hull-House wages maps, 1895.
Copyright Cornell University – PJ Mode Collection of Persuasive Cartography.

situated on alleys. The figures for the top two classes, green and yellow, are the mirror image of these results, with 70 per cent of street-facing buildings sitting in the top two classes (15.44 and 54.68, respectively) and only 12 per cent of alley buildings (8.96 and 2.83, respectively) classified as such. This spatial disposition of poverty and prosperity is not too distant from the marginal separation found in the previous chapter's analysis of London from the same period and shows how – at least west of the river – there was a diversity of class situated in what was viewed to be a poverty area. Given the descriptions of the quality of the buildings in the alleys, it is likely the differences between front and back were quite dramatic. Notably, the few buildings classified in the top classes that are present in alleys are situated on corners, so could in fact be reasonably assigned to the main road. Indeed, the notes on the map emphasise how the smart frontages on some of the main roads on this sheet were masking a 'hideousness shut up in the inside rooms of the larger, higher, and to the casual eye the better tenements of more pretentious aspect'.[19]

It is worth also bearing in mind the notation on the map regarding the alleys: they are 'opened by ordinance'. There was evidently a desire by the city to open up the most problematic of the densely built-up lots in a period prior to the drawing of the map. This is confirmed by the notes on the map, which describe how the rear tenements and alleys of the district contain 'the densest crowds of the most wretched and destitute . . .' (note the association of bad smells with disease in the following):

> Little idea can be given of the filthy and rotten tenements, the dingy courts and tumble-down sheds, the foul stables and dilapidated outhouses, the broken sewer-pipes, the piles of garbage fairly alive with diseased odours, and of the numbers of children filling every nook, working and playing in every room, eating and sleeping in every window-sill, pouring in and out of every door, and seeming literally to pave every scrap of yard . . .[20]

The decrepit state of the yards was quite shocking, and can be seen in the photographic collection of the Hull-House study, while a 1905 missionary tract attested to the miserable state of children in the 'slum district' of Chicago (see image from this missionary book in Figure 4.6),

> Look at that little weazened, half-starved girl of ten or a dozen summers down there at the foot of the stairway. This is a typical back yard in the slum district of Chicago . . . Her hair is dishevelled, her clothing is tattered and torn, her worn-out shoes but partially

Figure 4.6 A typical back yard in the slum district of Chicago, c.1905.
From Atkinson, *A Pathetic Case*, 1905; public domain via Wikimedia Commons.[21]

protect her feet from the stone pavement, her hands and face are grimy with the dust of the street.[22]

Overall, the Hull-House maps were intended to serve several purposes, to illustrate a method that could be used elsewhere, to present conditions to the public. The point was to stimulate inquiry and action, rather than offer solutions.[23] Together with the activities of the Settlement's residents, the impact of the maps continued well into the opening decade of the twentieth century.

Towards an academic sociology: Du Bois' map of the Seventh Ward of Philadelphia

A less-known successor to Booth's project was the map of the Seventh Ward of Philadelphia, drawn up by William Edward Burghardt (known as W.E.B.) Du Bois for his book, *The Philadelphia Negro*.[24] The study was commissioned by the Provost of the University of Pennsylvania in order to know 'precisely how this class of people live; what occupations they follow; from what occupations they are excluded; how many of their children go to school; and to ascertain every fact which will throw light on this social problem'.[25]

Du Bois' legacy as an urban sociologist is frequently omitted in the narrative arc that starts with Booth and continues with Hull-House and through the work of the Chicago School from the early twentieth century onwards. His pioneering study, although not novel in its conceptualisation of poverty (in effect, it built on the work of Booth and Rowntree), nor in its relatively limited coverage of a single district, was entirely original in its comprehensive and systematic empirical analysis of an area of Philadelphia's black population that until that time had not been considered worthy of analysis. Moreover, as Bulmer has argued, Du Bois' formal analysis constitutes one of the first sociological urban studies, which led him to develop original thinking about the influences of the social environment on poverty life in general and of minority groups in particular that were well ahead of their time.[26] The study was also the first step in a lifetime of scholarly work in sociology. Du Bois' innovation both as a sociologist and as a social theorist has frequently been marginalised, despite his being a pioneer in empirical sociological study methods that combine surveys, interviews and participant observation along with secondary data. The knowledge of statistics that he brought with him from his studies in Berlin is also frequently forgotten. In the case of his pioneering participant-observational study of Philadelphia, Du Bois made the most of his situation as a black American, to enter the 'Negro' [sic] area of Philadelphia, to live amongst the poor and to report on their life, providing a much closer insider's view than would otherwise be possible. There have been writers who have cavilled at this assessment, stating that as someone highly educated, Du Bois could not entirely have understood the realities of life in poverty. Nevertheless, he and his wife lived for a year in the area and experienced first-hand life in a run-down street at that time.

Du Bois had been born into a free black family in 1868 in a mixed area of Massachusetts, although he was amongst the few black people living there. After attending Fisk University in Nashville (funded by several worthies of his local community), he obtained a place at Harvard, which he funded through scholarships and his own savings from work. After completing his degree in record time, Du Bois went to pursue his PhD at the University of Berlin, taking courses in economic history and sociology for a thesis on the African slave trade. Martin Bulmer has written about the importance of Du Bois' time in Berlin, where he worked with Gustav Schmoller, whose research into plantation economics showed Du Bois how careful 'inductive analysis' using social scientific facts 'could produce systematic causal explanations of social phenomena'.[27] After the funding to support his PhD studies in Berlin ran out, Du Bois returned to the United States to complete his PhD at Harvard, although he never managed to get a post at an 'integrated' university and indeed his biography shows him to have lived quite an isolated life at Harvard, as one of the few black students there. In contrast with the examples of Booth, Rowntree and even the investigators at Hull-House – who all had independent means, or at least access to them – Du Bois lacked any external resources for his work. His scholarship relied on his own abilities to raise funds and the invitation to conduct the research in Philadelphia came at a time where he was teaching at an African Methodist church school in Ohio.

Du Bois wrote about how the study was instigated by one of the periodic attempts at political reform in the city, which wished to get to grips with what was seen as a corrupted vote emanating from the 'Negro Seventh Ward'. While commissioned by the University of Pennsylvania, his contribution to the study was never properly recognised by that institution, nor was it well supported financially. He writes:

> I was offered a salary of $800 for a limited period of one year. I was given no real academic standing, no office at the University, no official recognition of any kind; my name was even eventually omitted from the catalogue; I had no contact with students, and very little with members of the faculty, even in my department. With my bride of three months, I settled in one room over a cafeteria run by a College Settlement, in the worst part of the Seventh Ward. We lived there a year, in the midst of an atmosphere of dirt, drunkenness, poverty and crime. Murder sat on our doorsteps, police were our government, and philanthropy dropped in with periodic advice.[28]

The inquiry aimed to investigate the geographical distribution of the 'Negro' inhabitants of Philadelphia, their occupations and daily life, their home, their organisations, and above all, their relation to their million white fellow-citizens. Du Bois dedicated a whole chapter to describing 'The Problem', the 'peculiar social problems affecting the Negro people'. He describes this in terms of the lack of assimilation which, compared even with other unassimilated groups such as Jews or Italians is more entrenched due to its being bound up in problems of poverty, ignorance, crime and labour. For Du Bois, life in the 'Negro ghetto' was 'a city within a city;' its inhabitants did not 'form an integral part of the larger social group.'[29] In effect, Du Bois was arguing that black segregation was exceptional because it was comprised of multiple factors – social exclusion, as well as racial prejudice, as well as entrenched poverty – and all this coupled with immobility, a lack of opportunities to move out of the area. In conceiving of segregation as a spatial problem as well as a multivariate problem Du Bois' research predated spatial concepts of segregation laid out much later in the twentieth century.[30]

Du Bois was at pains to emphasise also the composition of the community: its stratification, with prosperous well-educated people as well as those at the other end of the spectrum living in relatively close quarters. Part of the purpose of his study was to add detail to existing knowledge of what was frequently seen as a homogeneous mass, with the majority blamed for the misdeeds of the small minority. His sample area was a particularly impoverished district of Philadelphia, but there was a black presence in many districts across the city. The study itself used schedules to ensure a consistency in data-gathering, which involved a house-to-house gathering of data over the course of a 15-month period. Du Bois distinguished between four groups:

> Grade 1. Families of undoubted respectability earning sufficient income to live well; not engaged in menial service of any kind; the wife engaged in no occupation save that of house-wife, except in a few cases where she had special employment at home. The children not compelled to be bread-winners, but found in school; the family living in a well-kept home. These were estimated to be eleven per cent of the Ward's black population.
>
> Grade 2. The respectable working-class; in comfortable circumstances, with a good home, and having steady remunerative work. The younger children in school. These were estimated to be 56 per cent of the Ward's black population.
>
> Grade 3. The poor; persons not earning enough to keep them at all times above want; honest, although not always energetic or thrifty, and with no touch of gross immorality or crime. Including the very poor, and the poor. These were estimated to be a little over thirty per cent of the Ward's black population.

Grade 4. The lowest class of criminals, prostitutes and loafers; the 'submerged tenth.' These were around six percent of the Ward's black population.

Du Bois' map, which coloured up the Ward based on these four social/income classes, did so by selecting only those lots with a black presence, leaving the rest blank (see Figures 4.7 and 4.8). The influence of Booth, both in the colour codes and in the labelling of the bottom class (as vicious and criminal) is obvious (and Booth is cited several times in his book), although the use of the lot as the unit of analysis follows Hull-House. By mapping his data, Du Bois was able to show the spatial patterning of the various strata of what might simplistically have been seen as the 'ghetto'. The very fact that the (at least potential) servant keepers were living alongside the servants themselves was a matter worthy of comment. Indeed, the nature of working in service is one of several topics discussed at length in the report. But this is not a simple matter: as Du Bois argued, the forces that kept the top class 'in the slums' – 'intermingled with . . . a dangerous criminal class' did not stem from a desire to retain proximity between the classes. Rather, it was factors such as discrimination (in employment, wages and housing) that led to cross-class proximity within the black population of the city, despite the fact that the 'Negro' population was *relatively* 'more scattered than ever before'.[31] Despite needing to live close to their place of employment, the city's black population was barred from doing so due to discrimination. Du Bois reports that if a black person wished to move out of an area, even if they were willing to rent the property, the putative landlord would charge a higher rent to make up for the possible loss of rental income from existing tenants moving out.

Aldon Morris has pointed out how Du Bois' view of the racial configuration of the black community in Philadelphia was as a planned, intentional programme, and not a 'natural ecological process'.[32] Yet is it also important to note Du Bois' careful reading of segregation being occasionally a matter of choice – he points out how black institutions, especially churches, served an important role within the community, which meant that they would act as a magnet for their congregants to remain close by; they would also serve as a refuge away from racial harassment. This aspect of racial and religious segregation will be discussed further in the next chapter.

Other matters are also worth further reflection. The lowest of the four classes is present in small specks of black throughout the area, but there

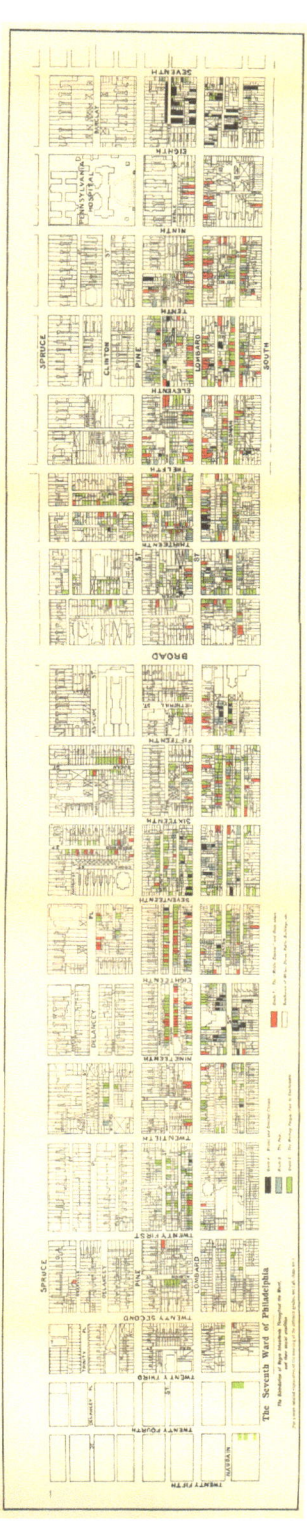

Figure 4.7 *The Seventh Ward of Philadelphia: The Distribution of Negro Inhabitants Throughout the Ward, and Their Social Condition*, 1896.

From the 1899 edition of Du Bois, *The Philadelphia Negro*. Image copyright University of Pennsylvania Archives.

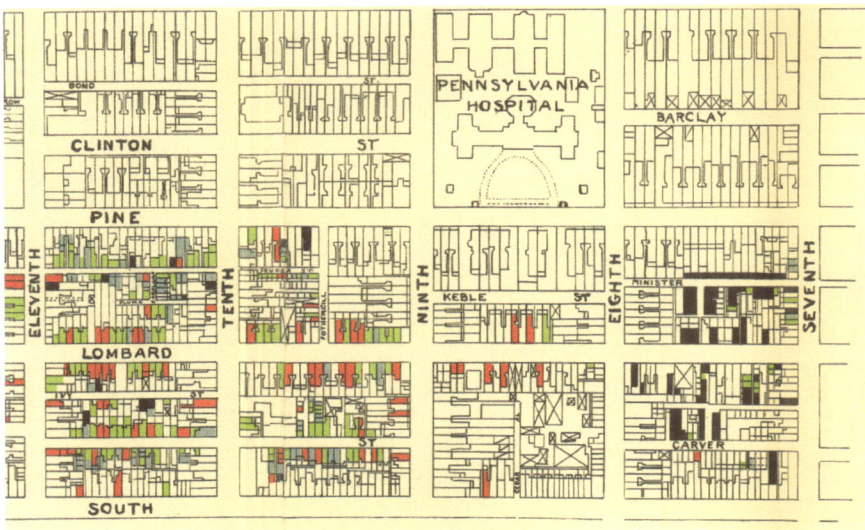

Figure 4.8 Detail, *The Seventh Ward of Philadelphia*.
Du Bois, 1896. Image copyright University of Pennsylvania Archives.

is a tight cluster of the criminal classes on the eastern edge of the ward, the site of the original black settlement in the city from 50 to 60 years earlier. Another aspect of the spatial distribution is the almost complete absence of the black population on the north–south wider avenues. They are clearly relegated to the relatively spatially segregated interstices of the district, with the lowest class located in the 'mostly crowded narrow courts and alleys', as Du Bois puts it. The local health inspector is quoted by Du Bois as assessing one of these areas (near Fifth and adjoining parts of Seventh):

> Few of the houses are underdrained, and if the closets have sewer connections the people are too careless to keep them in order. The streets and alleys are strewn with garbage, excepting immediately after the visit of the street cleaner. Penetrate into one of these houses and beyond into the back yard, if there is one (frequently there is not), and there will be found a pile of ashes, garbage and filth, the accumulation of the winter, perhaps of the whole year. In such heaps of refuse what disease germ may be breeding?[33]

In fact another spatial characteristic of the black households' distribution is apparent from analysis carried out by Amy Hillier in a recent GIS (Geographical Information Science) project on the Du Bois map, which mapped data from the 1900 census (the closest match to Du Bois' data period) on social class, race and national origin, household size, occupation of the head of household, presence of children, servants and boarders, and owner/renter status, as well as information about births and deaths from health registries.[34] The study confirmed empirically Du Bois' observation regarding a shift in demographics in the preceding 20 years since 15,000 migrants had arrived in the county from Maryland, Virginia and Carolina; Hillier shows that the black population can be considered as two separate groups from the point of view of their settlement patterns: those born in the city and those who had migrated into it from the southern states.

Du Bois commented of the rural incomers that, unlike the established population, they were untrained and much poorer, less habituated to urban living. They frequently ended up in the worse slums, living alongside the criminal minority. This is a feature of slum life about which Du Bois did not mince his words:

> The new immigrants usually settle in pretty well-defined localities in or near the slums . . . where there is a dangerous intermingling of good and bad elements fatal to growing children and unwholesome for adults. Such streets may be found in the Seventh Ward, between Tenth and Juniper streets, in parts of the Third and Fourth wards and in the Fourteenth and Fifteenth wards. This mingling swells the apparent size of many slum districts, and at the same time screens the real criminals. Investigators are often surprised in the worst districts to see red-handed criminals and good-hearted, hard-working, honest people living side by side in apparent harmony. Even when the new immigrants seek better districts, their low standard of living and careless appearance make them unwelcome to the better class of blacks and to the great mass of whites. Thus, they find themselves hemmed in between the slums and the decent sections, and they easily drift into the happy-go-lucky life of the lowest classes and rear young criminals for our jails.[35]

In a further indication of his subtle reading of social space, Du Bois found that the poorest districts in which the migrant black population settled also largely contained 'foreigners', recent immigrants from Russia and Poland – mostly Jewish. Du Bois' historical review of the ebbing and flowing of migration into the city points out that the latter tended to move in where the black population was moving out.

The subject of migration continued to interest Du Bois, who analysed it not only in *The Philadelphia Negro*, but in subsequent studies he conducted at the University of Atlanta, where he went on to work. The pioneering infographic on the shift in population from rural counties to the cities (see Figure 4.9) as well as their relative proportion in those cities is one of many such statistical graphics created by Du Bois for his later studies.

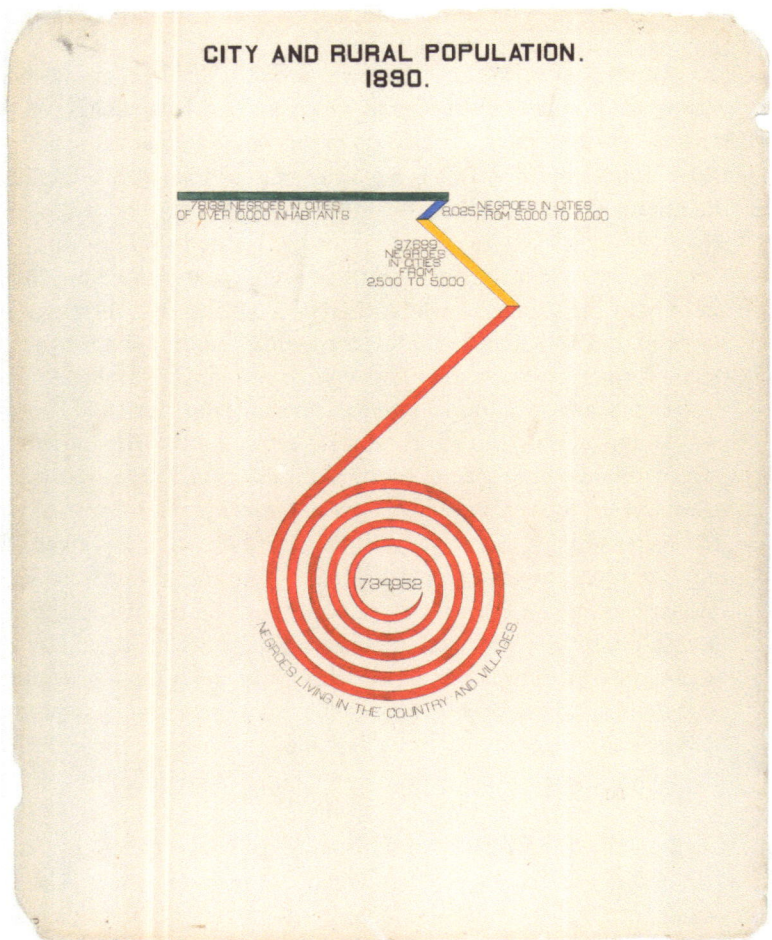

Figure 4.9 *City and Rural Population*, c.1890.

Chart prepared by W.E.B. Du Bois for the American section of the *Paris Exposition Universelle*, 1900. Daniel Murray collection (Library of Congress LOT 11931).

Soon after its publication, Du Bois' work in Philadelphia received high praise in a number of journals, which commended his painstaking methods and objective interpretation. However, they were less convinced by Du Bois' predictions regarding the potential for the future assimilation of the 'Negro' population. While there has been a transformation in the levels of prejudice reported by Du Bois, its long-term impact, its direct impact in shaping patterns of settlement in many cities in the United States, can be still seen today. Despite the excellence of Du Bois' work, he himself suffered prejudice such that, despite his Harvard PhD, the only post open to him was at the segregated University of Atlanta.

Du Bois' work is a landmark in social analysis: not only for his graphic innovation, but also for his achievement in setting a standard for a scientific approach to studying society that was taken up by scholars such as Robert Park and colleagues at the Chicago School of Sociology. Following the Philadelphia study, with the exception of the New Survey of London Life and Labour of 1929–31, thematic mapping became a rare form of graphic representation. Instead, there was a shift towards the contemporary method of sampling the population, which had the advantage of providing for national coverage of social analysis, despite the consequential loss in detail.

Revisiting Booth: Llewellyn Smith and the New Survey of London

By the 1930s, London had undergone something of a transformation from the period of Booth's survey. A Labour government had brought in many changes to improve living standards. At the same time, its economic geography had changed, with a massive growth at the urban periphery. New land uses, new technologies such as street lighting, new systems of sanitation, together with widespread slum clearances and new housing, all meant that the city was starting to be reconfigured, with migration to the suburbs both by the working classes and the more prosperous classes.[36] At the same time, alongside new factories in the suburban districts, small-scale production became prevalent:

> … the [New Survey of London] discovered "a high demand on skill" and innovation among the new occupation of "machine operator"

in small engineering workshops, among "specialised labourers" in wood-working, hand-cutters and designers in tailoring and leather work, printing, jewellery making and scientific and musical instruments. Apprenticeships were declining, but 23,000 were enrolled in Technical Schools within the County, and piece-workers could earn high wages for "steadiness and care" if not skill.[37]

Many jobs, especially in light industries, shifted outside of the city's original boundaries. A large increase in social housing, constructed by the London County Council, meant that much of the poverty found in Booth's time was starting to be diffused or eliminated. In parallel the country's economic recession in the late 1920s and early 1930s led to migration into London for work. Within the heart of the city, the older pattern of poor and rich living in a '"pre-modern", front street-back street' arrangement continued, due to the ongoing demands from industry for manual labourers as well as the 'army of service workers' who continued to serve London's wealthy population (Figure 4.10).[38]

Figure 4.10 View from a rooftop of the corner of Ocean Street and Masters Street, Stepney, London. Rows of houses, a shop and pedestrians are visible. 1937.
Copyright London Metropolitan Archives, City of London (*Collage*: the London Picture Archive, ref 119415).

The New Survey of London Life and Labour came at this juncture as a follow-up to Booth's study by a team at the London School of Economics.[39] Its director was Hubert Llewellyn Smith (a former assistant on Booth's *Inquiry*), who aimed to repeat Booth's methods with a street survey. He was joined by a team that included Arthur Lyon Bowley: Bowley not only designed the research, he also personally led an additional study of the poorest eastern boroughs, which comprised a house-to-house study of a sample of 12,000 working-class households.[40] Llewellyn Smith summarised his study's findings as follows:

> It is satisfactory to find that the level of poverty in East London is now only about one-third as high as in Charles Booth's time. It is much less satisfactory to learn that, in spite of this shrinkage, there are still more than a quarter of a million persons below the poverty line. When we consider how low and bare is the minimum of subsistence of the Booth poverty line, it is impossible to rest content with a condition of things under which one in ten in the Eastern Survey Area are living below this level. If the great progress towards extinction of poverty is matter [sic] for sincere congratulation and encouragement, the magnitude and gravity of what still remains give no ground for facile complacence, but rather for disquiet and searching of heart.[41]

Indeed, the study was well received – for its scope, but particularly for its attempt to revisit Booth's study to examine the impact of several decades of urban change, and specifically to address the question of whether poverty in London was increasing or decreasing. Sally Alexander describes how it showed a general rise in income, with a shorter working day and improved literacy.[42] The additional free time meant that people had more time to spend on leisure activities, which included the very popular cinema-going, as well as music – and gambling. She shows that access to more cash made for an easier life, but still there were a large number of people living in a state either close to or in poverty.

The nine volumes of Llewelyn Smith's report constituted a comprehensive follow-up to Booth's study 40 years earlier, with detailed studies on various aspects beyond the basic enquiry into poverty. Its accompanying set of six maps covered a wide area of London on a four-inch map of the County of London, coloured in a scale that reflected Booth's methods, though while black was retained for the 'lowest class of degraded or semi-criminal population', Booth's two other poverty classes were combined

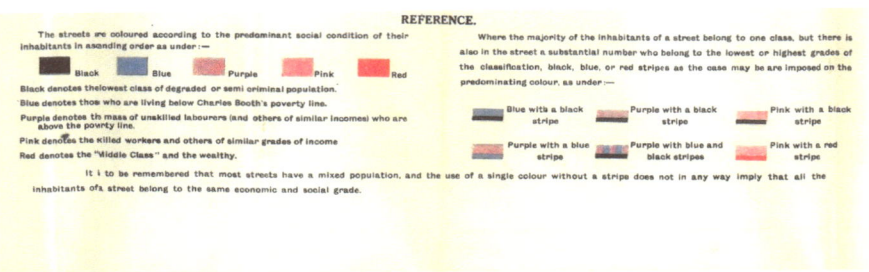

Figure 4.11 *The New Survey of London Life and Labour*; key to map.

H. Llewellyn Smith, *The New Survey of London Life and Labour: Survey of Social Conditions*. 9 volumes. (London: P. S. KING & SON, LTD., 1932), public domain.

into blue, denoting 'those who are living below Charles Booth's poverty line' (see Figure 4.11).

The report's definition of the poverty line has drawn some criticism. First, the street survey aimed to replicate Booth's methods, using school attendance officers, supposedly in the same vein as the School Visitors of Booth's study. However, the officers were unlikely to have any detailed knowledge of families with children, let alone those without (as E.P. Hennock puts it rather less politely, the survey and maps were 'conducted in the mental equivalent of historical fancy dress'.)[43] Similarly, Colin and Christine Linsley claim that the New Survey's poverty line failed to replicate Booth's survey as it included Class E, 'which was never a poverty class under Booth'.[44] They maintain that a more precise definition of poverty would have found that only 6 per cent of the population of London was in poverty, less than the 9.8 per cent calculated by Llewelyn Smith (and in contrast with Booth's 30 per cent). Despite this criticism, it is clear that Bowley's research design was undertaken with great care and based on a substantial amount of prior research and testing of methods in his own work.[45] The survey avoided defining a poverty line; rather, it defined a minimal standard of living. Bearing all this in mind, the survey maps are an important snapshot in time that provides a rare opportunity to look at aspects of spatial/social continuity and change. Nevertheless, for the purposes of reading them, it helps to be aware that the light blue may cover cases where people were living a degree above the poverty line defined by the survey.

One example useful in studying continuity and change is Portland Road in west London. Looking at it on the 1898–9 Booth map of poverty, one finds that the street bisects two districts situated to its east and west. This is due to the fact that the relatively smart Ladbroke Estate (to the west) was constructed alongside the much more downmarket Norland Estate, home at the time of its construction to piggeries and potteries, alongside a gypsy encampment (to the east).[46] The Booth map also illustrates a dramatic drop from the more prosperous southern section to the northern section, coloured light blue (poor) and darker shades of dark blue and black (vicious, semi-criminal) in the back streets beyond. By the time of the New Survey of 1929, the north end had become 'Degraded and Semi-Criminal' – coloured black for the lowest class (Figure 4.12, with inset showing the New Survey). Shortly afterwards the tenement housing was demolished and replaced by social housing: Notting Wood and Winterbourne House. Whilst the southern section slipped down the poverty scale into multiple occupancy, the council-owned housing locked in place a class situation, which essentially has not shifted to this day. The situation in the prosperous south has been dramatically different. It would probably be classed in the top grade if Booth were to repeat his survey today.

In a similar fashion, a study of the south-eastern sections of the New Survey maps shows that despite a general uplifting in poverty levels since Booth's time, the area around the docks is one of several instances

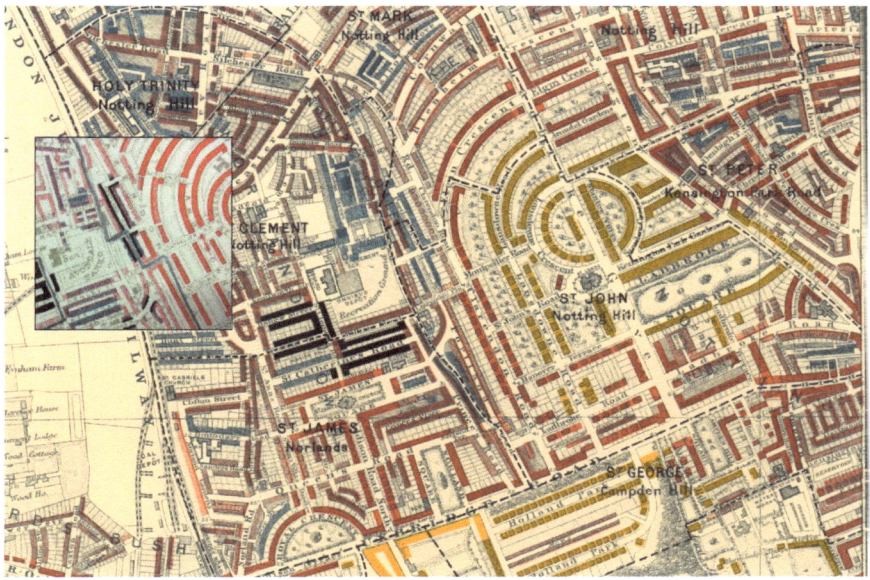

Figure 4.12 Detail of *Map Descriptive of London Poverty*, 1898–9, sheet 1, with inset of same area from *New Survey of London Life and Labour*, 1929–31.
Copyright London School of Economics Charles Booth Archive.
LSE reference no. BOOTH/E/1/8

of continuing problems with deep poverty, as can be seen in Figure 4.13, which has an inset of the same area as it was coloured up in the 1898–9 survey. So, while poverty has become more dispersed, it remained entrenched in some areas. This is confirmed by Brinley Thomas, who wrote soon after the survey's publication that while only one fifth of the 'Eastern Area' population were classed as poor in the New Survey, as opposed to over three-fifths in Booth's time, spatially, the segmentation of areas by industrial or transport infrastructure remained a problem:

> The corresponding proportion in Charles Booth's day was over three-fifths. The ugly pockets of chronic poverty still remaining are due chiefly to physical obstacles to freedom of movement. For example, "bounded by Limehouse Cut to the south, the gasworks to the west and a maze of railway lines to the east is the notoriously poor and degraded area formerly known as the 'Fenian Barracks'" which was the subject of a vivid description by Charles Booth and still maintains something of its old reputation.[47]

Figure 4.13 illustrates the area as it was assessed in both periods (with the main map showing the 1898–9 map coloured up following the above quoted remarks). It is clear how some of the streets within it remained impoverished well into the New Survey's period.

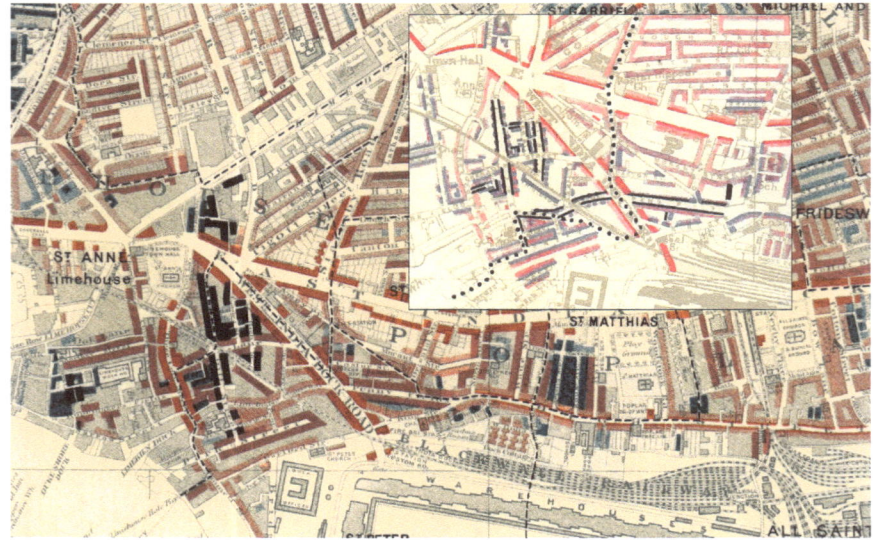

Figure 4.13 Detail of *Map Descriptive of London Poverty*, 1898–9, sheet 1, with inset of same area from *New Survey of London Life and Labour*, 1929–31.

LSE reference no. BOOTH/E/1/1.

In fact, George Duckworth's report on 'Fenian Barracks' (Fenian Barracks was the local name for Furze Street) was telling, not just because it revealed how the police viewed it as troublesome, an area where people they were pursuing could disappear from view with the collusion of their neighbours; but also from the point of view of the proposed solution to the problem, namely to reconnect one of the internal streets into Bow Common Lane so that there would be more through-traffic coming down it:

> This district has many bad spots and a great many very poor streets. From the police point of view the "Barracks" is the worst. This is a spot, which . . . is consistently bad, and from its size very difficult to deal with. Small streets may be bad in themselves but they can be tackled. A large block is another thing altogether.[48]

The spatial structure of poverty and – in this instance, crime – is a topic that will be discussed in greater detail in Chapter 6. Meanwhile, it is important to note that the spatial patterning of poverty continues to be a current topic of interest. Recent work by Oliver O'Brien has an interesting angle on this, where he proposes to construct a new way of visualising poverty that harks back to Charles Booth's methods, by overlaying UK national statistics on a map, showing demographic data according to census areas. Taking on board the limitations of choropleth maps, which tend to hide the detail of the street layout, he has created a visualisation which colours the streets across several major UK cities according to the demographic data of the area in which they are situated (Figure 4.14). Nevertheless (as O'Brien himself is at pains to point out), despite the graphic power and sophistication of the maps, they cannot replace the ability of the historical maps to distinguish between single houses (or even streets); the colour coding on an individual house is not necessarily strictly accurate.[49] This is one of many similar attempts to visualise the complexity of poverty in contemporary times. Other studies have used data on house sales to plot detailed data on housing economics, or land use patterns.[50]

The persistence of poverty – the continuing puzzle

The demise of poverty maps since the New Survey of the early 1930s was mirrored by the demise of the social survey during the inter-war years. Such surveys were seen as problematically detailed with too many facts

Figure 4.14 Limehouse area of London showing Index of Multiple Deprivation (2015) ratings.

Most Deprived decile in red, through orange and yellow through to light green and dark green for *Least Deprived* decile. Contains National Statistics and Ordnance Survey data © Crown copyright and database right 2011–15.
Image copyright Oliver O'Brien, Consumer Data Research Centre (CDRC).[51]

and figures, often delaying reform. Instead, the rise of ethnographic studies of communities meant that qualitative studies grew in popularity, while the development of sampling techniques, especially after the Second World War and the rise in professional marketing, led to an increase in the use of sample studies in urban sociology.

In many cities across the world the deterioration of inner-city dwellings continued to leave such areas stigmatised as slums, with many areas being written off for mortgage purposes. Entire tenements started to be torn down to be replaced by 'projects' – the new morphology featuring tall blocks of apartments, which in some instances replaced entire districts. In New York, for example, the mayor Fiorello La Guardia famously urged the clearance of slum areas with the slogan 'tear down the old, build up the new . . . Down with rotten antiquated rat holes . . . Let in the sun.' The projects solution was seen as entirely positive to start with, yet a combination of bad design, a lack of investment, and a concentration of social problems in a single area brought about rapid deterioration in many of these schemes. In the case of the Pruitt-Igoe project, for example, Mark David Major has shown that the scheme (constructed in 1954 in a predominantly black neighbourhood in north St. Louis) replaced a sophisticated grid layout comprising street-facing dwellings with a new, Modernist scheme that reoriented public space inwards, creating a chaotic layout with little spatial logic and dramatically lessened connections to the wider street grid. Although poor design did not determine the project's social deterioration directly, it is clear from Major's analysis how a combination of a spatial layout that made the project vulnerable to crime, the socially vulnerable population placed within it and its subsequent neglect by the authorities led collectively to a downward spiral of social malaise, and ultimately – though it should not have been inevitably – to the demolition of the buildings.[52]

There were similar cycles of deterioration in large-scale experimental projects elsewhere. In the UK, developments such as Sheffield's Park Hill Estate became symbolic of many of the subsequent social problems with inner city housing. This belied the fact that the widespread problems with the country's great Modernist programme of social housing can be blamed both on architectural and urban design defects and on a lack of investment, maintenance and social support.[53] As in the case of Pruitt-Igoe, detailed analysis of the spatial configuration of a number of these projects (or 'estates' as they are termed in the UK) shows that the spatial segregation of many of these schemes from the wider street network was an important factor in their decline: their external isolation and their internal over-articulation together created very low rates of internal pedestrian movement, meaning there was little natural interface between locals and strangers passing through the estates. This meant there was also little mixing between younger and older people and – importantly – between poorer and richer people, resulting in a vicious circle of social decline that was exacerbated by other factors, including stigma, physical decline and so on.[54]

This analysis points to the complexity of poverty. Social disadvantages can include the spatial concentration of inadequate housing, bad health and other cycles of areal economic deprivation, which can affect factors such as access to open space for recreation or for walking, increased vulnerability to crime and unequal access to jobs due to poor public transport, making it harder to hold down a job or to obtain training.

This analysis also points to the need to consider spatial patterns alongside the social problems of individuals: an area which is isolated is more likely to have a concentration of people living in poverty, as we saw in both the Booth maps and the New Survey maps. Yet this concentration is not only due to the area being hemmed in or cut off by the railway lines, it is not only a matter of physical segregation: a lack of social integration may be to do with prejudice, as we saw in the case of the Philadelphia map. This argument works also in reverse: an area that has become labelled as not worthy of investment (such as the case of late nineteenth-century Chicago) will create situations where existing dwellings deteriorate to such an extent that only those with the least choice, such as immigrants or the jobless, will end up living there. In addition, as Ruth Lupton has argued, the physical concentration of multiple forms of disadvantage, such as the lack of social ties, alongside population mix, reputation and so on, will shape area characteristics: 'For example, disadvantaged individuals in an isolated area will form one set of social relations, while disadvantaged individuals in a well-connected area may form another.'[55] An additional element at play has been the unforeseen outcome of state-sponsored housing, which leads to the concentration of poverty as an outcome of the better housing having been sold to the private sector, leaving clusters of people who are the least well-off in the worse areas of the city.[56]

While there are those who argue that patterns of behaviour, or a 'culture of poverty', are the main reasons for the sustenance of poverty over time,[57] this seems unlikely, given that the population in such areas tends to be highly mobile – in fact, too mobile, experiencing what is termed 'churn', to create a stable community. The spatial argument holds stronger still once we consider the study by Danny Dorling and colleagues on the persistence of poverty in London. His study has concluded that given that the city's population is likely to have been almost completely replaced over such a long period, the spatial structure of the area itself must be a contributory factor in the continuity of poverty in certain areas from the 1880s to this day.[58] In fact, the correlation between poverty and spatial segregation mentioned above points to there being two spatial processes involved: the area effect, alongside the effect of the finest-scale constitution of an area's streets. It could be said that a combination of the two effects together contributes to the worst problems of poverty.

Contemporary research has shown that the poorer population sectors are marginalised due to perceived or actual social barriers.[59] In fact, a recent large-scale study of the configuration of poverty in the US has found that there are two models for this: the first describes cities where the affluent are relatively concentrated and centralised, but also have higher than average interaction with the poor – examples include Boulder, Charleston, Portland, Seattle and Tampa. The second model applies in cities where there is a sharp divide between affluent and poverty populations, the latter of which are much more likely to be ethnic minorities – examples include Chicago, Los Angeles, Milwaukee, Philadelphia and Tucson.[60] In the UK, housing inequality can reinforce racial divides, especially amongst new immigrants from the developing world.[61] Ultimately, the spatial patterning of poverty is frequently intertwined with racial segregation, about which more in the next chapter.

Notes

1. J. London, *The People of the Abyss (2014 Edition with Original Photographic Plates; Introduction by Iain Sinclair)* (London: Tangerine Press, 2014; first published 1903), p. 59.
2. O'Day and Englander, *Mr. Charles Booth's Inquiry*, p. 23.
3. 'I am much indebted to Mr. Charles Booth and his associates for valuable suggestions given from time to time during the progress of this investigation. In a letter received from Mr. Booth, which is printed on p. 800, he shows the relation which exists between the York figures and those which he had obtained for London. It is unnecessary to point out the significance and importance of the facts which Mr. Booth thus brings out.' B.S. Rowntree, *Poverty: A Study of Town Life*, 2nd ed. (London: Macmillan, 1902, first published 1901), p. ix.
4. Rowntree, *Poverty*, p. 112.
5. Anne Kershen has pointed out that the earlier investigations by Mayhew into 'the plight of tailors in London' had similarly convinced him that, 'contrary to prevailing mid-Victorian belief, it was poverty which led to drunkenness, not the reverse'. This was because pubs operated as labour exchanges, with men waiting there to be recruited for work. Kershen, *Uniting the Tailors*, p. 5.
6. Rowntree, *Poverty*, pp. 129–30.
7. B. Harrison, 'Pubs,' in *The Victorian City: Images and Realities (Past and Present & Numbers of People)*, ed. H.J. Dyos and M. Wolff (London, Henley and Boston: Routledge & Kegan Paul, 1976).
8. K. Karimi, 'The Spatial Logic of Organic Cities in Iran and the United Kingdom', In *1st International Space Syntax Symposium*, ed. Major, M. D., L. Amorim and F. Dufaux, 05.01–05.17 (London: University College London, 1997). While Karimi's maps were from a slightly earlier period, the alignments of the main streets were broadly the same as those in Rowntree map.
9. Harrison, 'Pubs,' p. 169.
10. Rowntree, *Poverty*, pp. 311–12.
11. David Sibley contends that Hull-House was important also for its political radicalism, which included giving refuge to immigrants threatened with deportation and latterly meetings of union leaders, socialists (including Sidney and Beatrice Webb), leaders of the British Labour movement and the anarchist Kropotkin.
12. Residents of Hull-House – a Social Settlement, *Hull-House Maps and Papers: A Presentation of Nationalities and Wages in a Congested District of Chicago, Together with Comments and Essays on Problems Growing out of the Social Conditions* (New York: Thomas Cromwell, 1895), p. 10.
13. Residents of Hull-House, *Hull-House Maps and Papers*.
14. Residents of Hull-House, *Hull-House Maps and Papers*, p. 3.
15. Residents of Hull-House, *Hull-House Maps and Papers*, p. 10.

16. Residents of Hull-House, *Hull-House Maps and Papers*, pp. 143–4. The passage goes on to state that it encompasses a portion of the city which has on its western side 'the least adaptable of the foreign populations' and on its eastern side, unskilled labour and the most '"mal-adjusted" foreigners . . . rural Italians, in shambling wooden tenements; Russian Jews, whose two main resources are tailoring and peddling, quite incapable in general of applying themselves to severe manual labour or skilled trades . . . hopelessly unemployed in hard times.'
17. Residents of Hull-House, *Hull-House Maps and Papers*, p. 11. There was in fact a progression in abstraction in the study methods from Booth's initial house-to-house survey, from which he built his street-by-street statistical method, knowing that these were representative of the initial survey. In fact, the Hull-House method's reliance on personal interviews, rather than using reports by officials such as clergy, is more akin to Rowntree's method than that of Booth.
18. It should also be noted that the statistical method had its critics; a review of the papers and maps a few months after the study's publication criticises the method of calculating the wages as it did not take account of family size. E.G. Balch, 'Hull House Maps and Papers: Review of A Presentation of Nationalities and Wages in a Congested District of Chicago, Together with Comments and Essays on Problems Growing Out of the Social Conditions,' *Publications of the American Statistical Association* 4, no. 30 (1895), p. 202.
19. Residents of Hull-House, *Hull-House Maps and Papers*, p. 5.
20. Residents of Hull-House, *Hull-House Maps and Papers*, p. 5.
21. J.F. Atkinson, 'A Pathetic Case,' in *The Missionary Visitor*, ed. Brethren's General Missionary and Tract Committee (Elgin, Illinois: General Missionary and Tract Committee, 1905).
22. Atkinson, 'A Pathetic Case,' p. 46.
23. Residents of Hull-House, *Hull-House Maps and Papers*, p. 13.
24. W.E.B. Du Bois, *The Philadelphia Negro: A Social Study* (New York: Schocken Books, 1899).
25. Quoted in M. Marable, *W.E.B. Du Bois: Black Radical Democrat* (Boston: Twayne, 1986), p. 25.
26. M. Bulmer, K. Bales, and K. Kish Sklar, *The Social Survey in Historical Perspective, 1880-1940* (Cambridge: Cambridge University Press, 1991), p. 181. See also the recent book by Aldon Morris, who states that Du Bois was in effect the founder of the field of sociology, writing that, along with the Hull-House study, *The Philadelphia Negro* 'was one of the first empirically based scientific studies of American sociology', preceding the work of the Chicago School by two decades (see Chapter 6 for more on the Chicago School's history). A. Morris, *The Scholar Denied: W. E. B. Du Bois and the Birth of Modern Sociology* (Oakland: University of California Press, 2015), p. 68.
27. Bulmer, Bales and Kish Sklar, *The Social Survey in Historical Perspective*, p. 172.
28. W.E.B. Du Bois, 'My Evolving Program for Negro Freedom,' in *What the Negro Wants*, ed. R.W. Logan (Indiana: University of Notre Dame Press, 1944), p. 38.
29. Du Bois, 'My Evolving Program for Negro Freedom', p. 4. A recent study by Deena Varner puts a slightly different spin on Du Bois' conceptualisation of criminality, claiming that his analysis reflects his awareness of the relationship between the exploitation of blacks and their marginalisation – both social and spatial. D. Varner, 'Nineteenth Century Criminal Geography: W.E.B. Du Bois and the Pennsylvania Prison Society'. *Journal of Historical Geography*, 59 (2018): 15–26.
30. M. Pattillo, 'Race, Class, and Crime in the Redevelopment of American Cities,' in *American Economies*, ed. E. Boesenberg, R. Isensee and M. Klepper (Heidelberg: Universitatsverlag Winter, 2012). See more on the complexity of segregation in the next chapter (p. 159).
31. Du Bois, *The Philadelphia Negro*, p. 348.
32. Morris, *The Scholar Denied*, p. 49. Morris makes the distinction between Du Bois' analysis and the 'ecology' of cities (with the idea that ethnic areas evolve naturally) that was to be developed by the Chicago School.
33. Du Bois, *The Philadelphia Negro*, p. 306.
34. A. Hillier, 'Invitation to Mapping: How GIS Can Facilitate New Discoveries in Urban and Planning History,' *Journal of Planning History* 9, no. 2 (2010). See also the website at https://worldmap.harvard.edu/maps/8015.
35. Du Bois, *The Philadelphia Negro*, p. 82.
36. Though these led to greater class segmentation, as the railway companies offered cheaper working-men's fares on some lines at the expense of other, more expensive lines: J. Polasky, 'Transplanting and Rooting Workers in London and Brussels: A Comparative History,' *The Journal of Modern History* 73, no. 3 (2001).

37. S. Alexander, 'A New Civilization? London Surveyed 1928-1940s,' *History Workshop Journal* 64, no. 1 (2007), p. 307.
38. See Dennis, *Cities in Modernity*, p. 143.
39. According to Llewellyn Smith, the study was initiated by the London School of Economics but funded by a variety of philanthropic organisations, including the Laura Spelman Rockefeller Memorial, the London Parochial Charities Trustees, and several City of London companies. H. Llewellyn Smith, 'The New Survey of London Life and Labour,' *Journal of the Royal Statistical Society* 92, no. 4 (1929).
40. Much detail on the study is summarised in the notes by R. Bailey and A. Leith, *Computerising and Coding the New Survey of London Life and Labour* (Colchester: University of Essex, 1997).
41. Quoted in N.B. Dearle, 'Review: The New Survey of London Life and Labour. Volumes III and IV: Survey of Social Conditions. I. Eastern Area and Eastern Area Maps,' *The Economic Journal* 43, no. 170 (1933), p. 310. The author does not give a precise source for the quotation, though it is likely to be somewhere in the survey volumes that are the subject of Dearle's review.
42. Alexander, 'A New Civilization?'.
43. Bulmer, Bales and Kish Sklar, *The Social Survey in Historical Perspective*, p. 210.
44. C.A. Linsley and C.L. Linsley, 'Booth, Rowntree, and Llewelyn Smith: A Reassessment of Interwar Poverty,' *Economic History Review* XLVI, no. I (1993), p. 89. See also Alexander's criticism of the method, including the imprecision of measurement when income fluctuates dramatically day-by-day: 'The street survey, completed in a week, gave a snapshot measurement of a mobile and complex set of problems which could only be grasped through a study over time. The household sample (an invaluable, under-used source), as the NSL acknowledged, was not wholly reliable: often only the head of household's occupation was known to the School Attendance Officer, married women's employment was underestimated and household income varied with phases of the life-cycle, family size, seasonal and casual work. A household well provided for on a Sunday could be living on the breadline by Tuesday or Wednesday.' Alexander, 'A New Civilization?', p. 309.
45. Bowley had by this date published two well-received studies of poverty: *Livelihood and Poverty: A Study in the Economic Conditions of Working-class Households in Northampton, Warrington, Stanley, Bolton and Reading* in 1915 and its sequel, *Has Poverty Diminished?*, in 1926. Both were a refinement of Rowntree's methods used for estimating primary poverty in York at the turn of the century. Of particular note is Bowley's introduction of random sampling into Rowntree's methodology.
46. J. Bullman, N. Hegarty and B. Hill, *The Secret History of Our Streets – London: A Social History through the Houses and Streets We Live In* (BBC Books and Random House, 2012) shows that the Ladbroke Estate was always intended for prosperous, servant-keeping professionals; in fact, today, it houses the super-rich: bankers and foreign property investors.
47. B. Thomas, 'The New Survey of London Life and Labour,' *Economica* 3, no. 12 (1936), p. 464.
48. Booth, 'Poverty Series Survey Notebooks (Online Archive)', BOOTH/B/346, p. 59.
49. O. O'Brien, 'Geodemographics of Housing in Great Britain: A New Visualisation in the Style of Charles Booth,' http://vis.oobrien.com/booth/.
50. S. Law, 'Defining Street-Based Local Area and Measuring Its Effect on House Price Using a Hedonic Price Approach: The Case Study of Metropolitan London,' *Cities* 60 (2017); L. Narvaez, A. Penn and S. Griffiths, 'The Spatial Dimensions of Trade: From the Geography of Uses to the Architecture of Local Economies,' *A|Z ITU Journal of Faculty of Architecture* 11, no. 2 (2015).
51. The classifications shown are only applicable to the residential houses on the map, and represent an average across the local statistical area; therefore the classification colour on a house is not necessarily representative of that house (or even street).
52. M.D. Major, '"Excavating" Pruitt-Igoe,' in *11th International Space Syntax Symposium*, ed. T. Heitor et al. (Lisbon, Portugal: University of Lisbon, 2017). See more on social malaise, design and crime in Chapter 6.
53. J. Lowenfeld, 'Estate Regeneration in Practice: The Mozart Estate, Westminster, 1985–2004,' in *Twentieth Century Architecture 9: The Journal of the Twentieth Century Society - Housing the Twentieth Century Nation*, ed. E. Harwood and A. Powers (London: The Twentieth Century Society, 2008).
54. B. Hillier, 'Can Architecture Cause Social Malaise?,' in *Papers Given to the MRC Conference on Housing* (London: Unit for Architectural Studies, UCL, 1991).

55. Lupton, '"Neighbourhood Effects"', p. 5.
56. P. Spicker, 'Poverty and Depressed Estates: A Critique of *Utopia on Trial*,' *Housing Studies* 2, no. 4 (1987).
57. A. Ravetz, *Council Housing and Culture* (London: Routledge, 2001).
58. Dorling et al., 'The Ghost of Christmas Past.'
59. D. Massey and R. Denton, *American Apartheid: Segregation and the Making of the Underclass* (Cambridge, MA: Harvard University Press, 1993).
60. See full analysis in R.E. Dwyer, 'Poverty, Prosperity, and Place: The Shape of Class Segregation in the Age of Extremes,' *Social Problems* 57, no. 1 (2010).
61. D. Dorling, 'Housing and Identity: How Place Makes Race,' in *Better Housing Briefing* Paper 17 (London: Race Equality Foundation, 2011).

5
Nationalities, race and religion

> Especially at week-ends, coloured men come there from other parts of London for African-style food, for girls, and for relaxation in a neighbourhood where their different appearance does not make them objects of particular attention.[1]

The ghetto and the ethnic enclave

The German sociologist Georg Simmel wrote in 1903 about how the city functions as an alienating environment that is strikingly different from the village or the town: in the city the individual has to adjust to the 'metropolitan rhythm of events'.[2] This shift from the individuality of life in the village or town to the complexity of social life in the city was something with which the urbanists at the University of Chicago were deeply concerned; specifically, the way in which the urban environment reshapes social relations as well as patterns of interaction. In this context Louis Wirth, whose book *The Ghetto* was written at the university, examined the history of Jewish settlement patterns in Europe as well as their manifestation after migration to cities such as Chicago.[3] Wirth went to work alongside the leading sociologists Park and Burgess at the university, and his explanation of the nature of urbanism 'as a way of life' has shaped thinking about how cities allow for the segmentation of interests, groups and communities:

> Characteristically, urbanites meet one another in highly segmental roles. They are, to be sure, dependent upon more people for the satisfactions of their life-needs than are rural people and thus are associated with a greater number of organized groups, but they are less dependent upon particular persons, and their dependence

upon others is confined to a highly fractionalized aspect of the other's round of activity. This is essentially what is meant by saying that the city is characterized by secondary rather than primary contacts. The contacts of the city may indeed be face to face, but they are nevertheless impersonal, superficial, transitory, and segmental. The reserve, the indifference, and the blasé outlook which urbanites manifest in their relationships may thus be regarded as devices for immunizing themselves against the personal claims and expectations of others.[4]

This argument was part of Wirth's theory of how the city can be an instrument for integrating minority populations (although Wirth was also critical of the way in which the city can cause the fragmentation of family life). Nevertheless, this role that the city can play in distancing individual behaviour, in allowing for multiple identities to coexist, serves as a counterpoint to the typical framing of social segmentation as problematical. Self-segregation by choice is in fact a common feature in cities; arguably it stems from a natural preference to mix with people like oneself. It does not preclude social integration, given the fact that individuals can have many different social connections across the city.

However, due partly to the challenges in understanding the complexity of segregation, and partly to the problematic possibilities of its extreme manifestations, the subject continues to attract debate over its causes, structure and outcomes. Much of the discussion stems from a disciplinary divide between the social and the spatial sciences. There is limited engagement in the social sciences with the role of space in shaping segregation, while the role of mobility in overcoming segregation and the role of time in dissolving patterns of urban segregation are also frequently overlooked. Similarly, the spatial sciences tend to be more focused on patterns of settlement than the individual and societal motivations for those patterns.

One of the more debated concepts within urban segregation literature is the contemporary use of the term 'ghetto'.[5] Historically speaking, its first documented use was in sixteenth-century Venice. As we saw in this book's introduction, the Venice Ghetto has become the touchstone of discussions on segregation due to its historical importance as the first documented instance of forced enclosure. In addition to its centrality to the history of segregation in Europe, the Venice Ghetto serves as an ideal starting point for discussions regarding the spatial complexity of segregation.

The term 'ghetto' continued to be most commonly used to describe Jewish quarters until the twentieth century, although in the United States the legacy of African-American segregation has more recently given the term an exceptional resonance in both academic and popular literature, shifting away from its association with Jewish quarters (whether enforced or voluntary, and whether situated on the urban periphery or at its heart), towards the most extreme examples of racial and social divide in the United States in modern times.[6] Given that fact, it is surprising to find that areas labelled as 'ghettos' today are sometimes confusingly simply describing residential clusters of homogeneous groups or ethnic enclaves. The need for a clearer distinction between spatial descriptors such as *ghetto* and *ethnic enclave* is crucial and there are several scholars who emphasise this point: 'The ghetto is negative, the enclave is benign; the ghetto is forced, the enclave is voluntary; the ghetto is real, the enclave is symbolic; the ghetto is threatening, the enclave is touristic'.[7]

As with the term 'ghetto', one of the difficulties in the use of the term 'segregation' is its multiplicity of meaning. It can variously be used to describe exclusion due to poverty, to ethnic group membership, to physical separation, to economic deprivation or to occupational segmentation, among others. When used to describe one of these states, the other potential meanings are strung along as well. So, for example, the ethnic cluster is assumed also to be cut off physically from its surroundings. Reading the vast literature on the subject, one would conclude that the ideal urban formation would be a perfect mix of class or ethnic groups, but there is little evidence for determining exactly at what point separation by social group becomes problematic. Nor is it clear how one would arrive at the platonic ideal of social mixing. Although examples of more successful cases of spatial organisation by class or age exist, it is striking how much they are the result of specific social situations at a given point in history. So, for example, the apartments of nineteenth-century Paris organised class differentiation according to distance from the street (with the poorest living at the top), while in Berlin organisation was more likely to be by the depth of the urban block, and in London it would proceed turn by turn from main street to back street: marginal separation by linear integration.[8]

Segregation as a specifically urban phenomenon has been documented and researched for well over a century, stemming from a general concern with mass migration which coincided with the emergence of the new science of sociology. This chapter will show that cities are settlements unlike any other, not only in their scale, but also because they bring together the widest possible variety of people.[9] Indeed, one could argue that the essential role of cities is to bring together and to organise diversity.

This chapter looks at a series of maps that are in effect touchstones in the history of segregation, from late nineteenth-century San Francisco to Chicago, London and onwards. By doing so it aims to illustrate how the elusive concept of segregation can be defined with considerable precision, once its spatial aspects are measured. We will also see that examples of enforced division have occurred throughout history, from colonial enclosures across Asia to apartheid in South Africa. Many historical examples still endure, such as in the United States, where the racialised assessment of financial risk in the past continues to be etched on the ground in many of the country's major cities.

Moral panic in nineteenth-century Chinatown

This book's introductory chapter showed how the latter half of the nineteenth century featured an increase in international migration, leading to a proliferation of newspaper articles on the apparent 'foreignness' of urban slums in places such as Britain, the United States and Australia. In the case of San Francisco, Nayan Shah shows how from the 1850s onwards the hostility of the popular press was matched by the concern of politicians and health officials, for whom the sanitary conditions in the then largest Chinese quarter in the world were a matter of serious public health concern. He writes of salacious 'press coverage of public health inspections' in which 'reporters described the Chinatown labyrinth as hundreds of underground passageways connecting the filthy cellars and cramped "garrets" where Chinese men lived.'[10] The situation was not helped by a longstanding animosity between the city's indigenous population and its Chinese residents, who had initially arrived during the Gold Rush, but whose subsequent presence during the 1860s depression led to tensions between the city's white and Chinese population – and in particular, incoming migrants, who were seen by local labourers as direct competition for jobs. In addition, the high density of Chinese living patterns, with many living in dormitories in beds used in rotation, led to assertions that their actual way of life caused disease to proliferate. Eyewitness accounts cast the Chinese quarter's 'serpentine and subterranean passageways'[11] as a veritable labyrinth, one seen by the police as a severe challenge to public order in much the same way that Booth's policemen viewed parts of London a decade later (and with similarly intolerant language).

By the 1880s, a committee was established by the San Francisco Board of Supervisors 'on the Condition of the Chinese Quarter', which aimed to uncover the effects of Chinese immigration on the locality. The committee's report, describing the Chinese presence in almost biblical language as 'pestilent', included a folding map (see Figure 5.1), which appeared both in the San Francisco Municipal Report of 1884–85 and in Farwell's inflammatory text 'The Chinese at Home and Abroad'.[12] The map aims to cover the principal area of San Francisco's Chinatown, a total of 12 blocks bordered by California, Stockton, Broadway and Kearny Streets.

The emphasis of the map graphically is on vice, with a colour coding that illustrates the spatial patterning of the various activities that were of concern to the authorities. It shows 'General Chinese Occupancy' (tan), 'Gambling Houses' (pink), 'Chinese Prostitution' (green), 'Chinese Opium Resorts' (yellow), 'Chinese Joss Houses' (red) and 'White Prostitution' (blue).[13]

Closer study of the variously coloured land uses on the map sustains the impression of vice being concealed from view. It shows that 'Gambling' is consistently located either hidden from the street in rear yards or in buildings off alleyways. In fact, the report has an appendix listing all the 'Barricaded Gambling Dens', which shows how the study took care to capture the precise configuration of Chinatown's putative labyrinth, the better for ascertaining how to tidy it up and rid it of disease (see Figure 5.2).[14] The same is the case for the few 'Chinese Opium Resorts', which are uniformly hidden away from the street (Figure 5.3). The two are described vividly as a 'twin problem', the writer not holding back on his criticism in a passage rife with racist language and replete with terms intended to emphasise his revulsion at the manner of living in the quarter:

> The twin vices of gambling in its most defiant form, and the opium habit, they have not only firmly planted here for their own delectation and the gratification of the grosser passions, but they have succeeded in so spreading these vitiating evils as to have added thousands of proselytes to the practice of these vices from our own blood and race. The lowest possible form of prostitution – partaking

Figure 5.1
Official Map of Chinatown, San Francisco, 1885.

Image copyright Cartography Associates, 2000.

Figure 5.2 The Street of Gamblers (Ross Alley), San Francisco, California (1898).

The notable absence of women in the photograph is due to the Chinese Exclusion Act of 1882, which limited Chinese female migration. The law also forbade Chinese men to marry white women. Arnold Genthe, 1898.

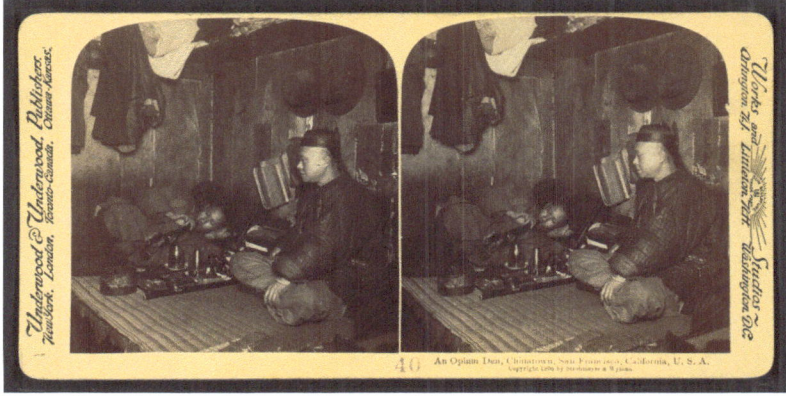

Figure 5.3 An opium den, Chinatown, San Francisco, California (c. 1900).

The fact that images such as these were available for sale is an indication of the exoticisation of the Chinese population at the time.
Robert N. Dennis collection of stereoscopic views.

of both slavery and prostitution – they have planted and fostered to a lusty growth among us, and have inoculated our youth not only with the virus of immorality in its most hideous form but have, through the same sources, physically poisoned the blood of thousands by the inoculation with diseases the most frightful that flesh is heir to, and furnishing posterity with a line of scrofulous and leprous victims that might better never have been born than to curse themselves and mankind at large with their contagious presence.[15]

'Chinese Prostitution' is located in more prominent thoroughfares and tightly clustered in certain areas: this may be due to the need to attract a certain amount of passing traffic. 'White Prostitution' (an especial horror from the point of view of the report, which stated that many of the women involved were opium addicts) is similarly patterned, but centred on the widest (and presumably most accessible) streets of the area. The map features a small quantity of buildings without colour, denoting 'White' businesses or dwellings (with the occasional 'Coloured' dwelling too). These are consistently located on the edges of the area.[16]

The report's text indicates how the mapping of Chinese 'Joss Houses' – namely Shen temples – was motivated by a highly negative reading of Chinese communal activity. The report's authors believed that the only solution to the problem of the Joss House was to convert the 'heathen' Chinese to Christianity (although ultimately the report recommends complete cessation of Chinese immigration):[17]

> . . . the 'Joss House' is, proportioned to population, even more common in Chinatown than are the edifices of the Christian church in other portions of the city. Idols of the most hideous form and feature squat upon their altars, from which license, in the belief of the Chinaman, sufficient to justify crime or vice of any degree may be had for the asking . . . Even the 'Goddess of Prostitution' sits enthroned upon her altar in more than one Joss House in San Francisco, and licenses her votaries to the practice of nameless indulgences and the most bestial gratification of their sensuous lusts. Let the sceptic who views this statement as an exaggeration or misrepresentation of fact visit the Joss Houses of San Francisco and he will no longer doubt; for it is the truth.[18]

Notably, of the thirteen Joss Houses identified on the map only three were on main roads. Eight were at least one turning off a main road, with an additional two either completely hidden or sharing accommodation with a business (in Duncome Alley and Stockton Place, respectively). The map provides a hint at the various spatial realisations of this minority community, with many of its places of worship being located in tucked-away corners of the district, similar to the location of the chapels constructed by London's immigrant French Huguenot immigrants who arrived predominantly in the late seventeenth and eighteenth centuries[19] or the synagogues built by Jewish immigrants in the eighteenth and nineteenth centuries.[20]

The Chinatown map's use of colour intentionally emphasises the most problematic aspects of the area's community in the eyes of the authorities. What is less obvious until one reads the map closely is how the uniform mass of buildings coloured in tan are actually comprised of an intricate mesh of businesses. Taking just one sample street (see Figure 5.4), while the first block of Sacramento Street leading south of Stockton Street is full of buildings labelled 'White Prostitution' (coloured blue), Dupont Street to its south is aligned with an array of retail

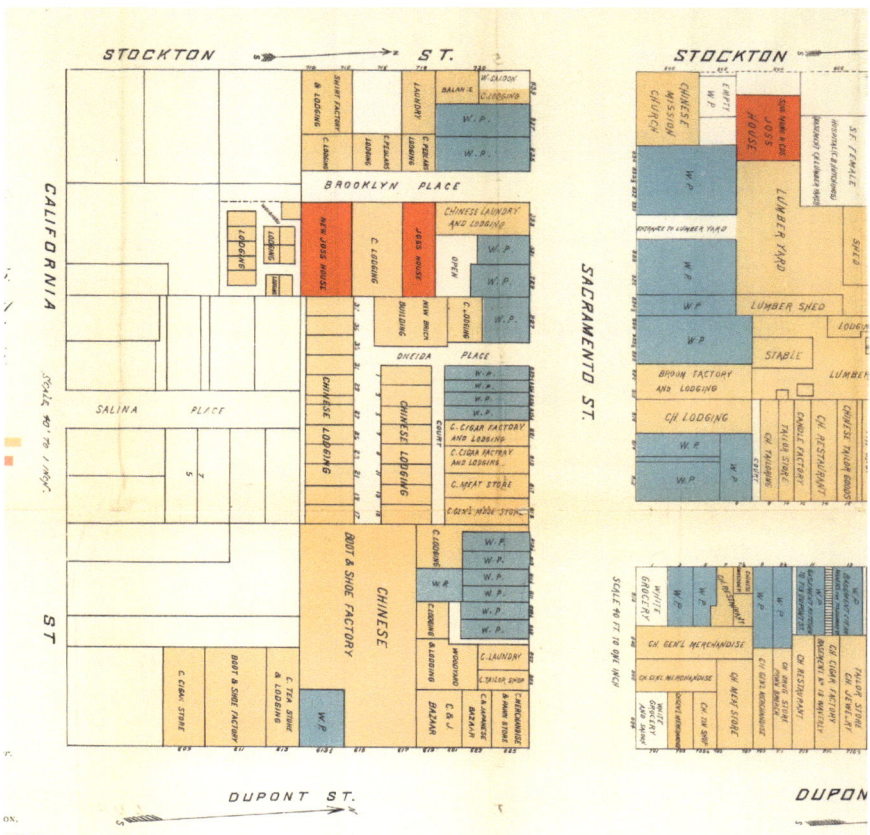

Figure 5.4 Detail of Fig. 5.1, *Official Map of Chinatown*, San Francisco, 1885.

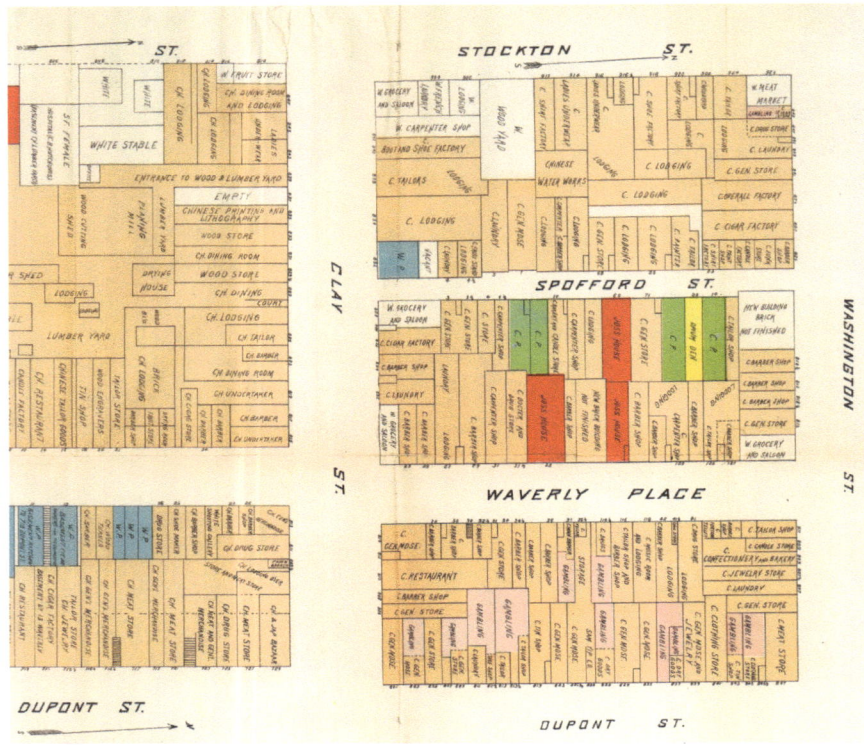

Figure 5.5 Detail of Fig. 5.1, *Official Map of Chinatown*, San Francisco, 1885.

and wholesale units: a meat market, a pawn and drugstore, a tea warehouse, shoe factory and many general shops selling merchandise. All are labelled Chinese. If we look at Figure 5.5, covering an area further east, and consider the businesses there, they are even more varied, with laundries, factories, shops, pawn brokers, bakery, restaurants and dining rooms; they present thus an entirely benign face to the main road, Washington Street.

This prompts a question about whether Chinatown was truly a place apart, or in fact formed part of the wider economy of the city. While the alleys were undoubtedly full of vice, and the Chinese community had its own commercial associations, it is clear that a certain degree of economic exchange must have occurred, despite the fact that overall the inhabitants of Chinatown were socially segregated from and politically marginalised by the city at large.

The map itself served its purpose. It captured in forensic detail the degradation of Chinatown's buildings along with its inhabitants. It played no small part in fuelling the continuation of racist challenges to the status quo of its pattern of living and, ultimately, helped support the authorities in bringing in the necessary public health regulations to tidy up its streets and alleys.

The spatial complexity of segregation: multiculturalism, nineteenth-century style

While Chinatown was being mapped in California, a more benign mapping of an immigrant quarter was taking place in Chicago. We saw in the previous chapter how the residents of the Hull-House settlement assisted in compiling the wage maps from data gathered, under the guidance of Florence Kelley, for a report commissioned by the United States Department of Labour, by order of the US Congress. The 1895 nationalities maps compiled at the same time constitute some of the most important maps of immigration in the United States (see Figure 5.6, showing maps 1–4). Through their level of detail, capturing the spatial patterning of nationality at the street-lot scale, they provided both for their time and for scholars of segregation today an astonishing picture of the way in which spatial clustering is formed locally within a district that might be otherwise simply considered as an 'immigrant quarter', or even a 'ghetto'.[21] They also reflect a sophistication in sociological method that preceded the scholarship for which Chicago became famous a decade or so later.

It should be noted that in her introduction to the 2007 edition of the papers, Rima Lunin Schultz criticises the almost uniformly bleak picture portrayed in the papers and maps, stating that if the boundary had been drawn so as to include the western side of Halsted, and some of the block beyond, it would have encompassed a more prosperous district containing a large French population.[22] This is undoubtedly true, and an important lesson can be learned from this on how the way in which areas are sampled can distort how they are evaluated. Nevertheless, putting aside the fact that the streets beyond were probably excluded due to the omission of those areas from the underlying maps, there is still much to be learned from these maps, which provide an unprecedented level of detail on the area at the time.

The maps' data were gathered at the same time as the wages information, between April and July 1893. The nationality of adults was recorded according to place of birth, while children under 10 took the nationality

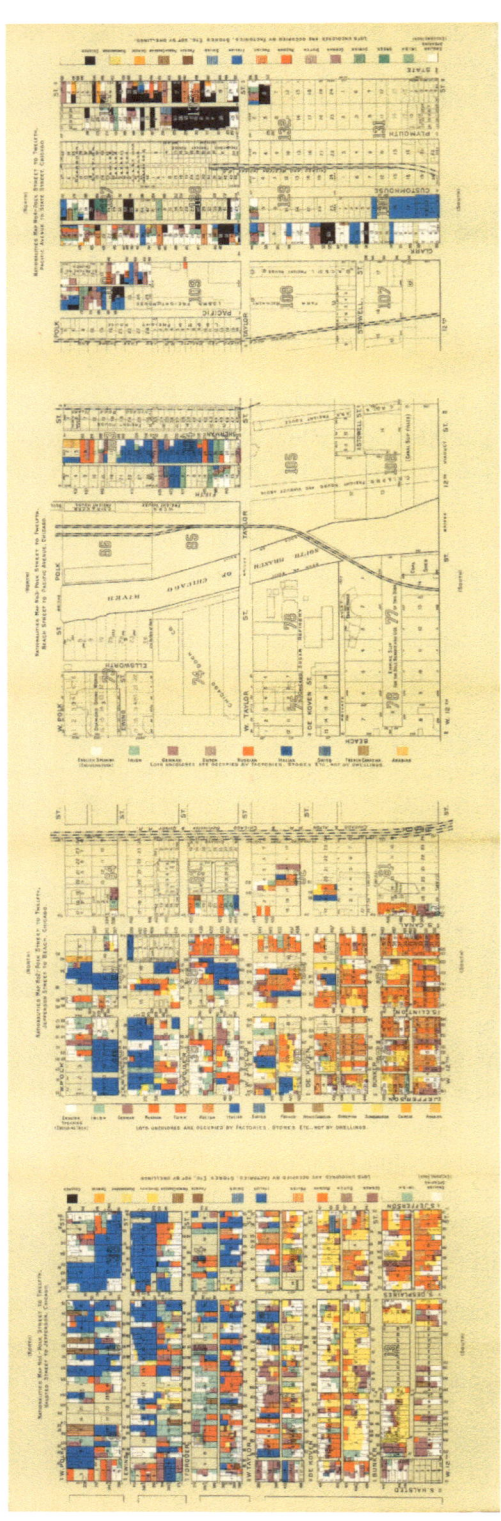

Figure 5.6 Hull-House nationalities maps 1–4, Chicago, 1895.

Copyright Cornell University – PJ Mode Collection of Persuasive Cartography.

of their parents. The map records 18 nationalities, using a colour code that ranges from black for 'coloured' to white for English-speaking. They show the percentage (not the sum) of each nationality residing on an individual city lot. The Irish are so distinct and important, state the map notes, that they are accorded their own colour: in line with the use of graphic stereotypes, it is green. Interestingly we have a distinction between 'Arabian' and 'Syrian', the latter which would include contemporary Lebanon. The notes inform the reader that the Russians and Poles are in fact 'uniformly Jewish'.

The resulting tapestry of colours shows the tendencies for nationalities to cluster, despite the overarching impression (at least in maps 1 and 2; see Figure 5.7) of a general intermingling of the nations into a single immigrant quarter. Further investigation finds the Italians (dark blue) solidly packed on Ewing and Polk Streets (in fact they are the most numerous in the district), while the Russian and Polish Jewish cluster is situated around Polk and Twelfth Streets. In fact, there was a much larger Jewish settlement that extended south from Twelfth Street, off the map. The maps notes inform us that a significant reason for the general pattern of localised clustering is the manner of sub-letting to boarders or lodgers to the rear of tenements, which frequently set in tow the 'prompt departure of all tenants of other nationalities who can manage to get quarters elsewhere'.[23]

This disposition for members of the same nationality to cluster together is a common feature of immigrant quarters that is frequently at the heart of what is negatively labelled as 'ghettoisation'. The same has been found in a study of Manchester and Leeds in the late nineteenth century, which found that between 60 and 80 per cent (respectively) of boarders and lodgers were living with families from the same country of origin; more still if only counting the Jewish quarter, rather than the city overall.[24] Cultural differences between immigrant groups would also cause them to choose to live in households of a common country of origin.

The same applies as well in patterns of settlement in contemporary cities. Pablo Mateos, for example, has found that in the neighbourhood of Kreuzberg in Berlin neighbours from different ethnic backgrounds mix successfully within the public realm as well as at school, yet at the scale of the apartment block, analysis of names on doorbells reveals a clear demarcation between blocks with a dominant Turkish (or other ethnic background) and those with a German-origin name.[25] This strong cultural reinforcement of place of origin can play a part in strengthening

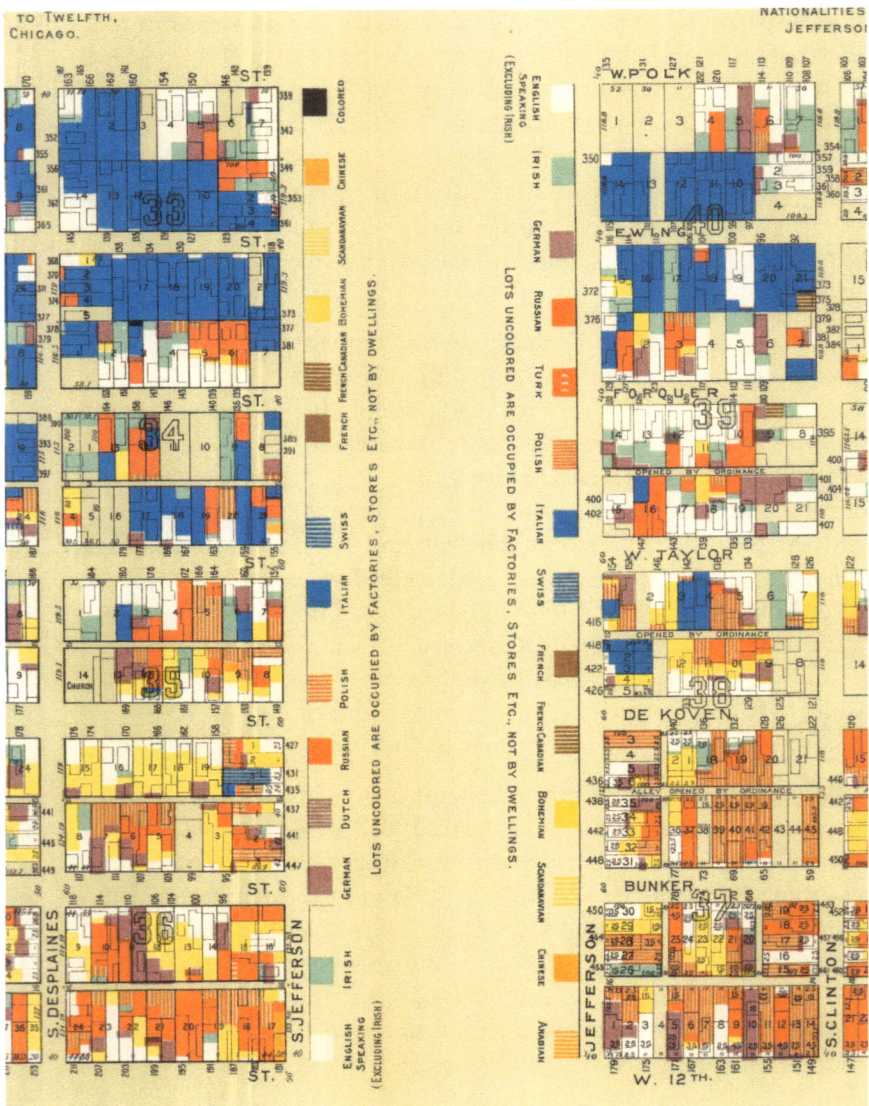

Figure 5.7 Detail of Hull-House nationalities maps 1–4, Chicago, 1895.
Copyright Cornell University – PJ Mode Collection of Persuasive Cartography.

communal ties and shows something important about how the scale of analysis can affect the way we read segregation on the ground. Taking Chicago as a whole, the area covered by the Hull-House maps would be seen as an immigrant quarter (with the tighter cluster of Jewish settlement that was known, as mentioned above, as the 'ghetto'), but within the area there is spatial segregation at building level. In effect, the map illustrates how geographical scales are socially constructed. They are the product of social relations, actions, and institutions. Separation at the apartment-block scale may be a completely benign outcome of patterns of ownership and housing economics.

The maps taken as a pair – wages and nationalities – allow the reader to cross-reference between levels of income and ethnicity. Even without detailed analysis, it is clear that poverty and nationality are intertwined, with a greater proportion of Bohemian and Polish immigrants located in the rear sections of lots and in alleys, while the eastern edge of the district had a disproportionate amount of black, poorer blocks (as discussed in the previous chapter).

The decade following the publication of the Hull-House maps yielded many other maps of nationality, whether directly influenced by its methods or as separate creations in their own right. In Boston a map 'Illustrating the Distribution of the Predominant Race Factors in the West End, Boston' showed the results of a study conducted by the Boston settlement headed by Robert Wood. Wood's study of 1903 resulted in two maps of the industrial character of the population and its racial composition. The map, which colours up street segments according to 'Americans; Irish [in green]; Jews; British & Provincials; Negroes [sic, in black]; Italians and Mixed' are said to be 'accurate as to the prevailing condition in each block'.[26] While no nationality is in a majority in the city, it is clear that there were localised clusters, though these were less obvious than in Chicago, both because of the use of the street as the unit of analysis and because the area studied was much smaller.

A decade or so later a much less precise approach to mapping nationalities was used for *A Map of Newark with Areas Where Different Nationalities Predominate*, illustrating the principal location of nationalities in Newark (Figure 5.8). The map, which was the frontispiece of a directory issued by the Bureau of Associated Charities, is mentioned only in passing in the directory. It does however feature alongside a detailed record of how cities such as Newark, with a population of 350,000 at the time, had formed a programme of private and public philanthropy to create a system of social services or agencies following rapid growth in the preceding years.[27]

With the Newark map, the mapping of race and nationalities in the United States arrived at a hiatus, as the focus shifted towards crime – although arguably crime mapping too was bound up in issues of race

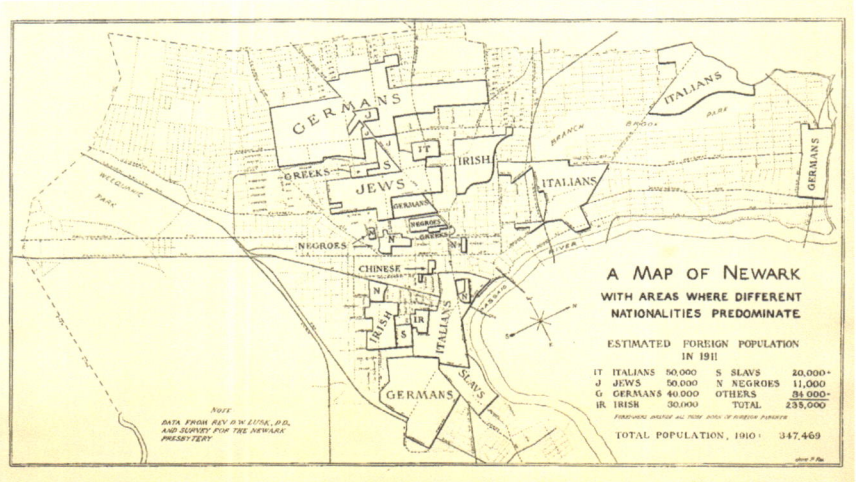

Figure 5.8 *A Map of Newark with Areas Where Different Nationalities Predominate.*
The resources for social service, charitable, civic, educational, religious, of Newark, New Jersey; a classified and descriptive directory.
Bureau of Associated Charities, Newark, N.J., and A. MacDougall, 1912. https://archive.org/details/resourcesforsoci00bure.

and nationalities, as we will see in the next chapter.[28] In this regard, the story of the racialised assessment of mortgage risk, what is known as redlining, will be considered later in this chapter, but meanwhile we need to return to London to consider one of the most important legacies of Charles Booth's project, a map of Jewish East London from the turn of the twentieth century.

Out of the ghetto? Jewish East London

Following a series of pogroms in the Russian Pale of Settlement from 1881 onwards, and the consequent loss of security concerning both life and livelihoods (whether directly or indirectly shaped by the ongoing reduction in circumstances for an already impoverished population), a large influx of refugee Jewish immigrants started to arrive in London and other major English ports.[29] Some were en route to the Americas while others remained in England. The consequence was that the Jewish population of major English cities grew rapidly in the latter two decades of the nineteenth century, with London's swelling to around 135,000.

The map of Jewish East London (Figure 5.9) appeared in *The Jew in London*, published in 1901 in response to widespread concern about

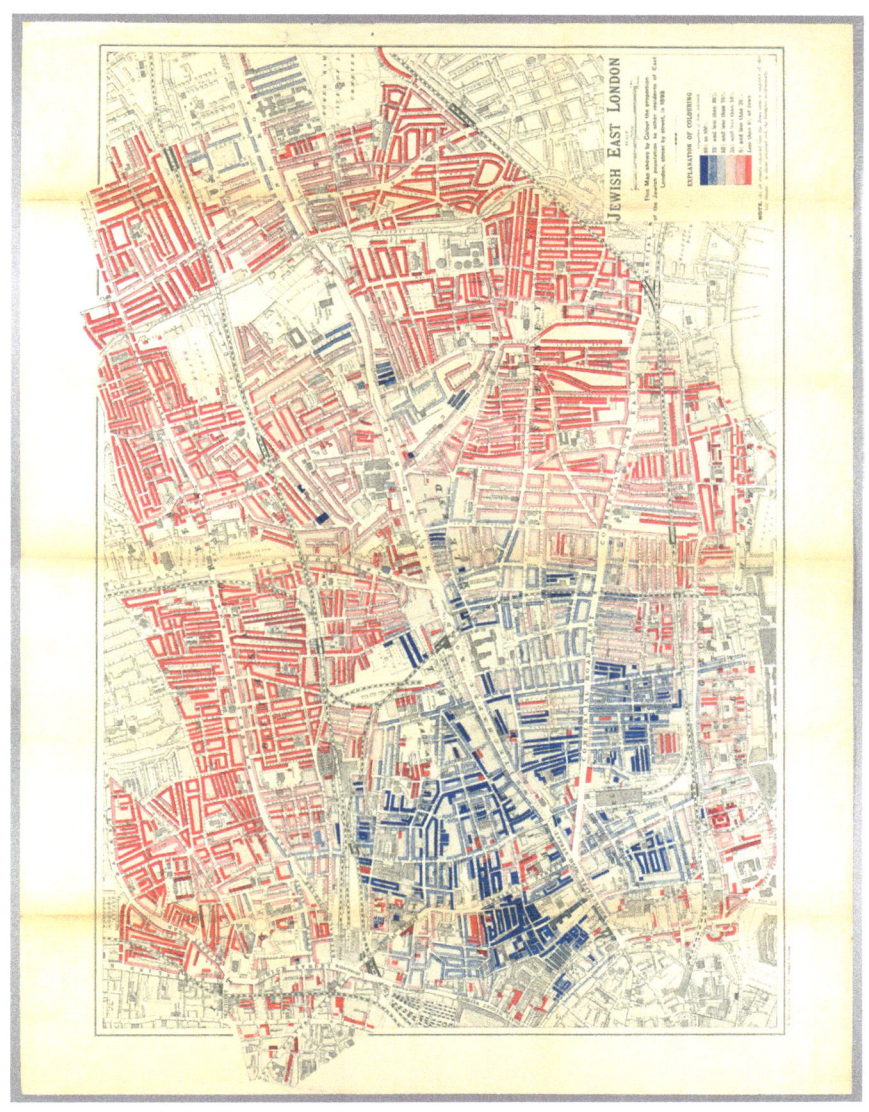

Figure 5.9 *Jewish East London*, 1899.

George Arkell, 1901. Copyright Cornell University – PJ Mode Collection of Persuasive Cartography.

the incoming migration of the preceding decade and a half. The book, produced under the auspices of Toynbee Hall, contained two essays: one by Charles Russell, who wrote as an outsider; the other by Harry Samuel Lewis, a member of the Jewish community.[30]

Numerous Jewish organisations had by this point been set up by the established Jewish community to provide charitable support, but also with the aim of integrating the arrivals socially and economically into the existing population (partly out of pure charitable instinct, but also to avert anti-Semitic responses). Despite this aid, the problems of high-density settlement included crises concerning unsanitary conditions and overcrowding alongside rent inflation. Jerry White describes how 'it was said that rents in the Jewish quarter had nearly doubled – at a time when wage rates were rising only slowly'.[31] Immigrant living conditions were frequently worse than those of the other inhabitants of the poverty areas, but this did not help the impression that the newcomers were the cause of a general deterioration in living conditions in the area. Booth states that, putting aside the generally bad conditions in some areas of the East End of London, the Jewish quarter was distinctive in featuring 'overcrowding in all its forms, whether in the close packing of human beings within four walls, or in the filling up of every available building space with dwellings and workshops. . .'. Housing density was higher in this area than anywhere else in the East End.[32] The book's writing, reasonably measured (for its time), recommends that instead of restricting immigration the focus of legislation should be on remedying the worst evils: the 'sweating system', which could be deterred by encouraging larger factories; and domestic overcrowding, which could be mitigated by encouraging dispersal farther afield, to cheaper accommodation. It also argues that English-born Jews should join local clubs and show that they were willing to join in with the 'great nation' to which they now belonged.[33]

The map of Jewish East London was drawn up by George Arkell, one of Booth's investigators. Arkell obtained data from the London School Board as well as Tower Hamlets and Hackney's school visitors. As in the Booth study itself, Arkell used the visitors' schedules, which contained information on all families with children of school age (that is, under 14). Data on every street in the area provided information on which families were Jewish and which were non-Jewish, identifying Jewish families by name, the school attended and whether Jewish holidays were observed. As in the Booth study, the School Board Visitors' data were extrapolated, with the assumption that in any given street the proportion of Jewish

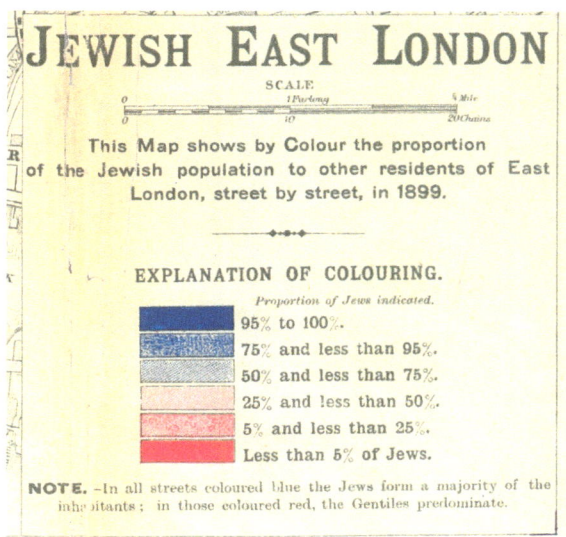

Figure 5.10 Key to map of *Jewish East London*, 1899.
George Arkell, 1901. Copyright Cornell University – PJ Mode Collection of Persuasive Cartography.

and non-Jewish families without children under 14 would match that for families with school-age children. The map was then coloured up, with the street as the unit except in cases of very long streets, which were instead dealt with in sections (see key to map in Figure 5.10).[34]

Close study of the map reveals that in fact there was widespread intermingling of Jewish and other residents on the streets of the area, with most streets in the middling range of mixing. The pattern of settlement is reflected in the book's map notes, which give a hint at how the street's spatial configuration interacts with the way in which incoming migrants had settled in the area over time:

> The gradual spread of the Jews . . . has followed . . . the path of least resistance. From Whitechapel the flow has moved along the great highways, especially Whitechapel Road and Commercial Road, and into the streets immediately off these thoroughfares. In streets not directly connected with the main roads, and not readily reached, the influx has been slow and is comparatively recent. In some long streets directly connected with a main road, a distinct difference may be noted between the near and remote ends of the street. . . The same tendency to spread along the main thoroughfares is seen in the outlying portions. . .[35]

Figure 5.11 Detail from Charles Booth, *Maps Descriptive of London Poverty*, 1898–9.

Showing poverty classes ranging from red (middle class) down to black ('vicious, semi-criminal') Booth, *Life and Labour.*

Historical evidence indicates as well that in addition to work, it was the availability of cheap housing that made the East End attractive to impoverished incomers. However, this was not the only factor in the spatial distribution of Jewish immigrants. As we saw in Chapter 3, 'the poor were not a homogeneous class'.[36] The fact that some streets were more accessible than others was not unknown to the incomers to the area. While the newest arrivals had no choice but to lodge in whatever cheap and inaccessible accommodation was available, those that managed to improve their economic situation, although this might take years, made the most of the spatial logic of the area to move into the streets with greater accessibility and, concomitantly, a greater intermingling of Jewish and non-Jewish people.[37]

Opportunities for spatial integration were of benefit to the economic activities of the incoming migrants. Indeed, there were Jewish traders who regularly travelled beyond the reaches of the East End. Booth describes, in one of many examples, the area around Chalton Street, situated perpendicular to the busy thoroughfare of the Euston

Road: 'Jews come from Whitechapel, selling draperies for the most part' at a daily market, which was busiest on the Friday (the eve of the Jewish Sabbath).[38]

Even at the finest scale, small shifts in the street geometry can interrelate with the social situation: this can be seen in Figure 5.11, which zooms in on the same area as featured on the 1898–9 Booth map and the Jewish East London map in Figure 5.12. The corresponding area is described in the following passage from Booth's police interview notebooks:

> West along Brushfield Street, north up Gun Street very rough. Mixture of dwelling houses and factories. Three storey and attic houses. A Jewish common lodging at the north-west end. Where the Jew thieves congregate. It is called "the poor Jews home" [sic] on the board outside. South of Brushfield Street Gun Lane is rougher than the north end. Street narrow. Loft across from wall to wall. Old boots and mess in street. 4.5 storied houses, a lodging house

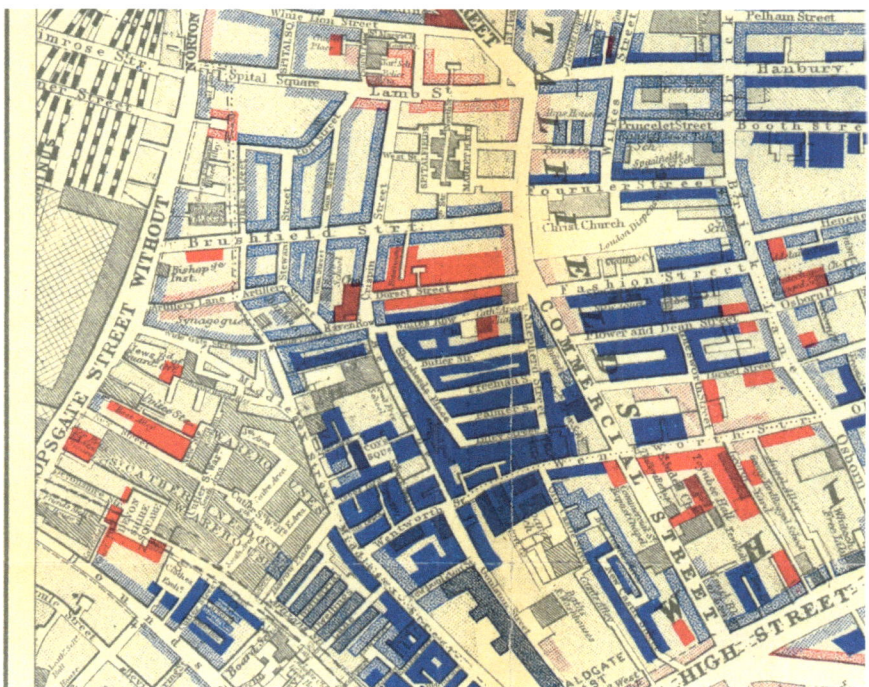

Figure 5.12 Detail from *Map of Jewish East London*, 1899. Streets with a Jewish majority are shown in blues and streets with a Jewish minority in reds.

George Arkell, 1901. Copyright Cornell University – PJ Mode Collection of Persuasive Cartography.

at south east end. Dilapidated looking: ticket-of-leave men living here... But "it is not a street particularly noted for prostitutes!" At the north end is Fort Street. Fairly well to do. Pink rather than purple of map: "Jew middlemen live here"... Steward Street 4 storied. Windows dirty but pink - in map purple. Duke Street has houses on east side. The west side is all factories and warehouses. Character dark blue to light blue. In map purple. "The coster flower & fruit sellers in Liverpool Street come from here!" Inhabitants are a mixture of Jews and Irish. South into Artillery Lane. Three storied synagogue on west side. Dwelling houses on east side only. Purple to pink. West along Artillery Passage all Jews. Rather narrow passage with shops on either side. Pink as map. On the north side of it is a passage leading to Artillery Lane called Artillery Court not coloured on map, ragged children, fish curers, rough, dark blue. East along Artillery Lane past the Roman Catholic dormitory at the corner of Bell Lane. The hour was only 1PM but there was already a crowd of 30 men and 2 women waiting to be taken in, though the doors do not open till 4. French [the policeman] said there were a set of scoundrels but they did not look as if they belonged to the worst class, all fairly clothed, one or two old cripples.[39]

The notebook extract details how Artillery Passage is lined with shops and is relatively prosperous (indeed it is of the same character today). It is likely that it benefits from connecting directly to Bishopsgate, the main artery – an ancient road – that runs north–south immediately to the west of the district. The notebook description also shows that, while the different class and religious groups might be separated at the residential scale (in the back streets), once in the busy main roads they were not only 'co-present' – the basic ingredient of community – but also had the potential for social interaction with the host society, at the very least through trade and industry, but also through a network of social interdependence and support, with a large number of communal charitable and religious organisations set around the area.

The map of Jewish East London is in fact a reminder to read maps with caution. On the face of it, the map is a neutral record of the statistical picture of Jewish immigrant settlement at the time, showing this population's spread along the main roads and the streets adjacent to these, with certain areas remaining completely empty of the newcomers. Yet, the choice of the colder colour, blue, to denote streets with a majority presence, in contrast with red for streets where Jewish inhabitation was in a minority, gives the impression that the Jewish presence in the area was much greater – and much more problematic – than it was in reality. The map also excluded pockets of Jewish settlement in the more prosperous districts of Dalston and Hackney. Nevertheless, the map stands as an important record of the heart of Jewish working-class settlement in London at this point in time.

In contrast, the Booth police interview notebooks attest to the intermingling in the same street of Jewish and non-Jewish inhabitants, the latter in reality largely Irish, many of them children of migrants from a previous generation. There is other historical evidence for mixing across the classes and indeed across supposed ethnic and spatial divisions, where work for members of the Irish community was found more easily amongst the Jewish new immigrants than from the community at large.[40] In effect, as Bronwen Walter has written, the lives of the two communities 'intersected at three nested scales – the household, the street and the wider sector of the East End'.[41]

It should be noted though that other sources report on tensions between the communities: 'in those districts on the edge of the foreign quarter, where street supremacy had not been settled, resistance to Jewish encroachment was most intense . . . [and] led to the formation of Jewish exclusion zones'.[42] However, whether it was the case that people moved to seek amenable neighbours or for the simple reason of cost, there may equally be instances, here as elsewhere, where immigrants chose to cluster for cultural or religious reasons.

Strong rules against intermarriage have historically created clusters of Jewish settlement beyond the initial stages of migration, and south-east Asian immigrants to the UK have upheld similar rules in order to maintain cultural cohesion and occasionally to avoid contact with 'what they see as a prejudiced host society'.[43] The need for a group to have a sufficient presence to maintain its religious institutions frequently explains immigrant communities' remaining in an area beyond the first stages of settlement. Russell and Lewis found this to be the case in Jewish East London, where individuals seemed 'often to remain in the district, out of regard to the feelings of their parents, who are perhaps dependent on them for support',[44] or because of the presence of Jewish institutions, especially synagogues, in the district.

The synagogues were one of the prominent land uses in the area. They tended to be located on secondary streets, not facing the main public streets, but taking advantage of local routes used by the community.

Indeed, in a study where over a hundred Jewish institutions were plotted on a map of the area, it was found that other than the synagogues, all other Jewish community institutions, including clubs, schools, theatres, soup kitchens and colleges, were located on streets which were spatially integrated in relation to the local street network. Of these, the streets with educational institutions and streets with more than one institution type had average local integration values which were significantly higher than the average for the whole area.

More recent analysis compared the visibility of synagogues and churches in a section of the East End at the turn of the twentieth century.[45] One of the cases studied was Chevrah Shass Synagogue, an image of which can be seen in Figure 5.13. If we look at its location on the section of the Goad Plan in Figure 5.14 (where it is marked as 'Old Montague Street Synagogue'), we see that although its entrance was visible to the street, the synagogue itself was almost entirely hidden from view, tucked away

Figure 5.13 Chevrah Shass Synagogue, Old Montague Street, East London c.1950. The synagogue entrance was marked by a sign in Hebrew and English c.1946–59.

Artist: John Gay; copyright Heritage Image Partnership Ltd/Alamy.

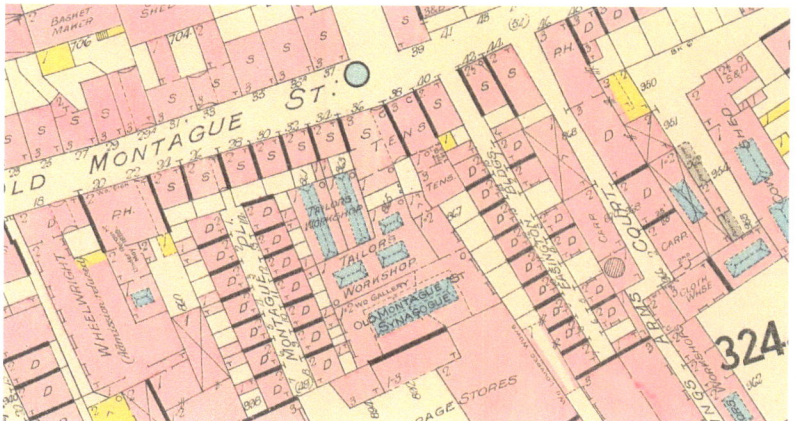

Figure 5.14 Detail of Goad Plan, sheet 322, 1899, showing location of Old Montague Street Synagogue (Chevrah Shass).

Crown copyright and Landmark Information Group.

at the end of a passageway between two shops and behind factory stores. It was one of 14 synagogues within a small area of Whitechapel. Overall the study found that the Jewish inhabitants of the district had a variety of prayer spaces, ranging from ad hoc prayers that took place in workshops through more formalised (typically back-yard) buildings to a cohort of synagogues that were conversions from chapels, with only a couple of purpose-built structures. The analysis found that while most of the synagogues were either completely hidden, or only visible to their immediate surroundings, the two most visible synagogues were situated on streets serving the more longstanding Jewish residents of Whitechapel.[46]

At the same time, Jewish communal institutions were placed in prime positions within the principal streets local to the neighbourhood, leaving the most outward-looking economic activities to take place on those streets that could best benefit from London's natural flows of movement. The study concluded that the incomers were able to take advantage of the fine-grain street system of London's East End to construct several complementary networks – economic, social and cultural – to sustain their community during their first stages of acculturation into wider society.

Redlining, Apartheid and the persistence of segregation

The maps of nationalities, race and religion shown so far have varied in the explicitness of their discussions of race, with the San Francisco map being the most extreme in tone. Nevertheless, all three have in common the fact that they were recording cases where a minority was settling in a cluster more or less by choice (if one puts aside restrictions due to poverty, or localised racism). In contrast, the maps discussed in the following section are a sample of the many cases worldwide where there has been explicit seclusion of a minority group due to government-inspired racial laws.

Carl Nightingale has revealed how as early as 1711 the city of Madras was divided into a 'White Town' and 'Black Town', while J.A. Schalk's *Plan of the City of Calcutta, 1823* shows the Black Town to the north standing in contrast with the widely spaced compounds of the White Town to the east of the city's fort, 'carved out of the jungles of Chowringhee'.[47] Although segregation by race was never completely successful, it remained a constant feature of twentieth-century colonial planning. It can be seen too in many cities across Africa, such as Asmara, for which a 'racial zoning map' (see Figure 5.15) was drawn when the country was under Italian rule. The map is an overlay on the first town plan for the city, which was drafted in 1913.[48] We can see how in drawing up the plan, its architects anticipated a future European-style layout of boulevards, squares and public gardens set within radial roads.

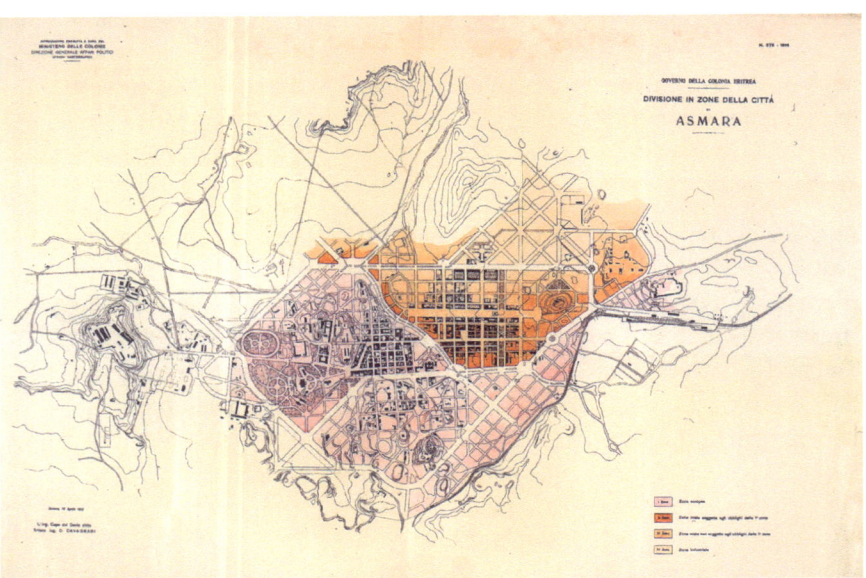

Figure 5.15 *Racial Zoning Map of the City of Asmara*, 1916.
Courtesy Dr Edward Denison and the Asmara Heritage Project.

The map was based on a new building code that had been issued to form the basis of future zoning that required a racial separation – that is, segregation – of the city into 'four distinct urban quarters'. The zones comprised one for Europeans only, a second in which Europeans and other foreigners (Jewish, Greek and Arab merchants) would mix with Eritreans working in the market, a third for natives, located in the area surrounding the Orthodox Church, and a fourth reserved for industry. Belula Tecle-Misghina points out that unlike many other colonial cities, Asmara's plan took account of pre-existing physical elements in the landscape as well as social conditions, and that these together helped to shape the allocation of the areas in question.[49] She states that 'in effect, Asmara's development was founded on a curious amalgamation of pre-existing social and physical realities overlaid by a racially predisposed socio-economic policy of development'.[50] Similarly, Denison and colleagues have stated how the plan reflects the tensions between 'ethnic and religious diversity' that were 'characteristic of many modern colonial encounters'.[51] What seems evident from reading the map is that the drive to shape inter-racial encounters (or, more precisely, to avoid them), took primacy in deciding on the zoning: almost from its inception the 'native' zone was disconnected from the European zone, situated beyond the industrial zone and linked with just two boulevards. It is interesting however to note that as the city grew, the zoning took secondary position to the overarching urban design characteristics of the city, namely its 'picturesque, grid, and radial elements'.[52] At least in this instance, Modernist planning principles were stronger than social structures, a phenomenon possibly assisted by the many phases of political change that the city experienced throughout the twentieth century.

Following his study of black Philadelphia (which we saw in Chapter 4), W.E.B. Du Bois prophesied that 'the greatest problem of the twentieth century would be the problem of the colour-line'.[53] He maintained that this barrier to integration would persist so long as there was a lack of contact across the racial divide. This would stem from a variety of factors, including a lack of physical proximity (due to the way neighbourhoods are organised); a lack of economic opportunity; a lack of political power; and less tangible contact, either through an absence of opportunity for the intellectual exchange of ideas or an absence of daily social contact or religious teaching:

> First, as to physical dwelling. It is usually possible to draw in nearly every Southern community a physical colour-line on the map,

on the one side of which whites dwell and on the other Negroes. The winding and intricacy of the geographical colour-line varies, of course, in different communities. I know some towns where a straight line drawn through the middle of the main street separates nine-tenths of the whites from nine-tenths of the blacks. In other towns the older settlement of whites has been encircled by a broad band of blacks; in still other cases little settlements or nuclei of blacks have sprung up amid surrounding whites. Usually in cities each street has its distinctive colour, and only now and then do the colours meet in close proximity. Even in the country something of this segregation is manifest in the smaller areas, and of course in the larger phenomena of the Black Belt.[54]

In this context, the 'redline' maps drawn in United States in the 1930s provide an ideal historical record of the spatial division Du Bois alludes to. The practice of redlining started when the Federal Home Owners' Loan Corporation, a national source of credit for companies, set out in 1933 to demarcate the relative risk of loans for mortgages in different areas of cities across the United States. The corporation's local assessors ranked neighbourhoods according to four grades: A to D, colour coded as green, blue, yellow and red, respectively. They considered factors such as

> intensity of the sale and rental demand; percentage of home ownership; age and type of building; economic stability of area; social status of the population; sufficiency of public utilities, accessibility of schools, churches and business centres; transportation methods; topography of the area; and the restrictions set up to protect the neighbourhood.[55]

In reality, almost all black neighbourhoods were classified as grade D, red – hence the term redlining (in fact, red shading would be more precise). Notably some of the maps had formal categories typed in, while others, such as that of Birmingham, Atlanta, cut to the chase, with the categories green to red labelled 'Best, Still Desirable, Definitely Redlining, Hazardous' respectively, with an additional grey shade for 'Negro Concentrations' (see Figures 5.16 and 5.17).

The financial implications of a bad grading were severe, as most loan companies and insurers would refuse to lend money in redlined areas. In addition, by condemning entire areas as being at high risk for defaulting, the maps became self-fulfilling prophesies on the future of such districts

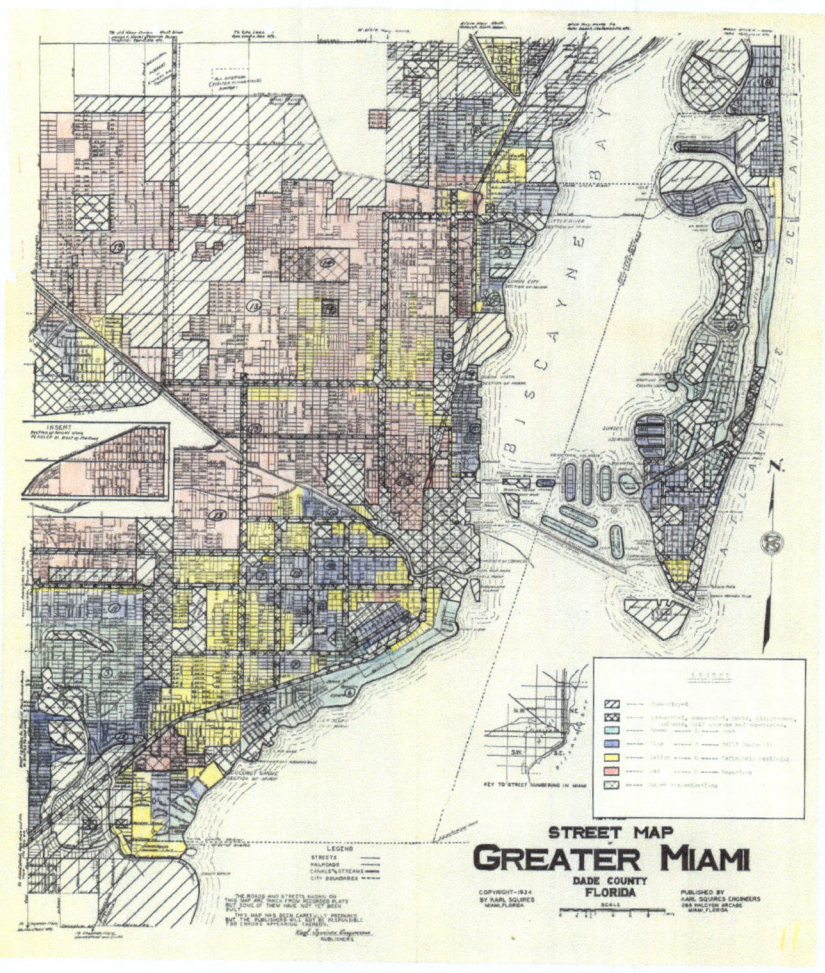

Figure 5.16 Redline map of Miami, Florida, created by the Federal Home Loan Bank Board.

Home Owners' Loan Corporation, 1933–1 July 1939; Series: Residential Security Maps, 1933–1939, Record Group 195: Records of the Federal Home Loan Bank Board, 1933 –1989. Source: https://catalog.archives.gov/id/6082409.

as poverty areas – a phenomenon that, as we saw in the previous chapter, had group effects, such as a lack of access to resources, to training and to work.

There is evidence that in the mid-twentieth century there were cases of racial discrimination in the United States from estate agents as well as from potential neighbours of African Americans wishing to move into

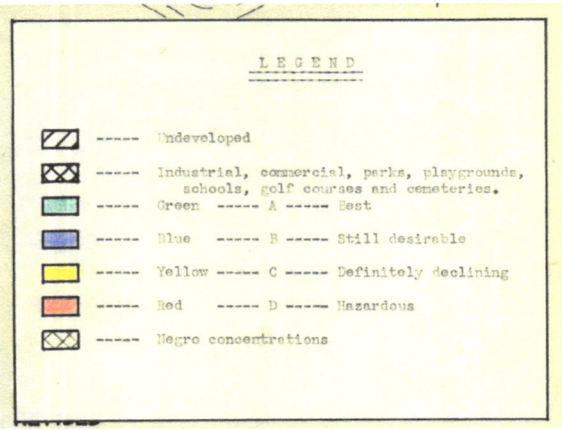

Figure 5.17 Key to Miami redline map.
Home Owners' Loan Corporation, 1933–1 July 1939.

new neighbourhoods, including both intimidation and violence. The impact of redlining on discrimination was different; it had the effect of areal discrimination, rather than prejudice against individuals or families. As Paul Bennett has argued, one of the reasons that the spatiality of redlining is so abhorrent is because people were: 'assumed to be poor risk and excluded because of their location of residence (and the racial composition of that area)', even if as individuals they did not at all represent a financial risk.[56]

The spatiality of redlining is even more problematic if we consider how it had the effect of setting in stone what might previously have been more fluid boundaries. A study by Amy Hillier shows the long-term outcome of such area-based discrimination in the case of Philadelphia.[57] Her study involved a random sampling of mortgage loans, mapped according to types of lender (federal, non-federal, commercial banks and so on). The study found that overall while people living in the central core of the city – where there was a predominance of poorer neighbourhoods with greater concentrations of blacks – had access to mortgages, they had fewer choices of lenders and had to pay a higher interest rate than people living elsewhere in the city. Although some middle-class blacks moved out of the city's core (due to their having greater opportunities to obtain cheap mortgages to do so), poorer people remained stuck in inner-city areas. Thus, inner-city districts where large numbers of people struggled economically in the past continue to struggle today, remaining the most marginal both economically and racially. Hillier discusses whether the persistence of poverty in the inner city can be explained by a simple rational response of lenders 'to the increased cost of doing business in a particular area',[58] yet she shows evidence that racial targeting remains a problem today in certain districts, even after controlling for the income of the applicant and the underwriter's evaluation of risks.[59] Either way, the outcome is the ongoing concentration of poverty areas that suffer from a lack of ethnic mixing – a pattern that can be found in many other US cities besides Philadelphia, such as San Francisco.[60]

There are even starker patterns of spatial inequality in the case of South Africa in the post-Apartheid era, following what was arguably the most extreme form of geographical confinement in the twentieth century. While many racial barriers to residential moves within the country have been removed via government legislation, there is plenty of evidence that longstanding problems of poverty in the country have created a situation in which many moves towards area integration are stymied by entrenched problems of poverty and racism that will be difficult to shift in the short term.[61]

In his masterly study of Chicago, *Great American City,* Robert J. Sampson found that the persistence of segregation in particular areas of the city was the result of three principal factors: low economic status, ethnic heterogeneity and residential instability.[62] Sources of 'compromised wellbeing' such as poverty and unemployment, which would normally be viewed as problems relating to individuals (that is, not as influenced by their environment), are shown to be consistently clustered. Sampson repeatedly shows how the multiple factors that comprise deprivation are interrelated socially and endure spatially. He finds that housing inequalities and racial stratification are 'etched in space' and proposes that the neighbourhood effect has a structural logic comprising social problems, exacerbated by racial segregation and the 'poverty trap', that together explain the spatial inscription of inequality. Evidently, without understanding the spatiality of segregation, such problems are likely to remain entrenched.

The complexity of segregation

One of the most important lessons to be learned from maps of segregation is the complexity of segregation as a socio-spatial phenomenon. While maps will typically capture a single aspect, such as race or nationality, it is clear that economics, politics and – fundamentally – urban space itself will all have a role to play in shaping how an individual, as part of a group, will be able to overcome their place in the city. As soon as we start to consider also the racial segregation of areas, especially when it is shaped by external factors (such as governmental restrictions), the many possible interactions between the various factors that make up the concept of segregation become complicated even further in a process that has been termed 'wicked'.[63]

This is one of the reasons that segregation is such an enduring phenomenon. While there are scholars who have produced mathematical models that claim to forecast patterns of segregation, they have to contend with the almost impossible challenge in predicting societal behaviour. After all, we cannot extrapolate from the actions of individuals to the likelihood that a group will behave in a particular way.

Another aspect of the challenge of capturing patterns of segregation are the issues of scale and space, highlighted above in the discussion around the spatial patterning of nationalities in Chicago. Whether the city was at that time segregated or not seems to be a question of scale, just as much as of space. Families congregated in certain areas of the district as the result of a sequence of individual decisions that collectively resulted in the patchwork of group clustering seen on the map.

The many ways in which maps can reveal spatial patterns means that their interpretation is going to depend on how a map maker chooses to represent the data they had gathered. We have seen how in the case of the studies of Chicago and Philadelphia the *building* or the *lot* were chosen as the unit of analysis. The result tended to correspond to this unit size: most solutions proposed in these cases were more likely to be social than spatial, quite possibly as a result of the spatial pattern being less obvious to the untrained eye. In London, on the other hand, Booth chose to use the *street* as his unit of analysis (other than the initial house-to-house survey). While this was primarily a pragmatic decision that stemmed from the vast scale of his project, nevertheless, although his maps were incredibly detailed, and successfully demonstrated the tractability of the poverty problem in London, they still gave the impression that poverty might be prevalent across a whole street, when it may in fact have been more localised. The proposed solutions were consequentially primarily spatial, such as slum clearance or dispersal to the suburbs.

This chapter has also demonstrated that the complexity of segregation as a concept means that what is frequently termed a 'ghetto' is not necessarily a place hosting a single ethnic group. Richard Sennett describes in *The Uses of Disorder* how a walk down Halstead Street, Chicago, at the time of Hull-House would take the individual past a vast range of people from many different countries, carrying out many different activities.[64] This diversity of uses belies the perception of the 'ghetto' as being an enclosed, spatially contained area.

Indeed, the spatial analysis of nineteenth-century London revealed that despite the label of 'ghetto',[65] the separation between poor and more prosperous streets in the East End of the city was not as sharp as perceived. Although there were pockets of severe deprivation, other poverty streets were located in close proximity to more affluent areas and there were distinct advantages to this mixing of classes, especially for commerce. Taking Middlesex Street (also known as Petticoat Lane) as an example, it is evident that the street's market, which had become predominantly Jewish at the time of Booth's study, benefited from its proximity to more prosperous streets in its vicinity. It managed to serve simultaneously as a place of work for the poorer inhabitants and a place of leisure and exchange for the more prosperous people of the neighbourhood; similarly, this was a place where immigrants and more established inhabitants were able to co-exist in the public realm:

> The neighbourhood of old Petticoat Lane on Sunday is one of the wonders of London, a medley of strange sights, strange sounds, and strange smells. Streets crowded so as to be thoroughfares no longer, and lined with a double or treble row of hand-barrows, set fast with empty cases, so as to assume the guise of market stalls. . . Those who have something showy, noisily push their trade, while the modest merit of the utterly cheap makes its silent appeal from the lower stalls, on which are to be found a heterogeneous collection of such things as cotton sheeting, American cloth for furniture covers, old clothes, worn-out boots, damaged lamps, chipped china shepherdesses, rusty locks, and rubbish indescribable. . . Other stalls supply daily wants – fish is sold in large quantities– vegetables and fruit – queer cakes and outlandish bread. In nearly all cases the Jew is the seller, and the Gentile the buyer; Petticoat Lane is the exchange of the Jew, but the lounge of the Christian.[66]

The mapping of diversity clearly has its ambiguities, but of course so can the underlying data. It is important to bear in mind, for example, that conceptions of foreignness, ethnicity and race are not fixed, but subject to the social or political context within which they were created. For example, the status of people born in Ireland was very much a social matter in the nineteenth century. Ireland had become part of the United Kingdom in 1801, but people of Irish origin were still considered outsiders by many people well into the twentieth century.[67] This dichotomy between official political status and everyday reality was even more complicated in the nineteenth-century USA, where stratification by 'colour' started as early as the 1870 census, 'which distinguished between "white," "mulatto," "Chinese" and "Indian"; by 1890 additional categories included "quadroon," "octoroon" and "Japanese".[68] These difficulties of definition have not gone away. In France, collecting statistics on an individual's race or ethnic identity is illegal, while in Sweden, there is a lack of detailed ethnic and racial statistics.[69] Although such policies have good intentions, the result is that policy-makers find it harder to check if there is discrimination in employment recruitment, or to tailor their social assistance to the specific needs of individual groups; nor can they easily unpick social problems stemming from poverty from problems relating to minority status (assuming the two differ).

Other problems with classifying the children of immigrants can raise thorny questions around what makes for a native – whether language, culture or even diet. Anne Kershen, for one, has written about how new arrivals to London's East End (whether eighteenth-century French Huguenots, nineteenth-century Russian Jews or twentieth-century Bengalis),

> used their mother tongue (or dialect) as a verbal building brick in the construction of a spatial location away from a previous home, a dwelling place where they could set down roots and accommodate change in an alien society. At the same time, they sought to create a fortress within which they could exclude all that was strange and threatening.[70]

Further down the line of acculturation, food may be the aspect of immigrant culture that people hold on to longest, even if they have abandoned traditional dress, language or music.[71]

Spatial integration is therefore much more complex than a simple map can capture. A group can start to disperse across a city's area, yet still maintain its core religious or cultural activities in the original place of arrival. Integration is also an aspect of time. People can have a residence in one place, but their workplace in another. A map of the same people at the same point in history will reveal strikingly different patterns depending on whether it is recording the former or the latter.[72] Similarly, the spatial pattern of an immigrant quarter can shift across the day and the week. For example, in a study of London's contemporary Chinatown, Simone Chung shows that its streets mix different people together at certain times, but also separate the various groups using the street at other times. On the face of it, a land use map of the area will show the area to be dominated by Chinese-owned or at least Chinese culturally related businesses, restaurants and other services, but a map of people's presence on the street reveals a different, more complicated picture. At some points during the week the map will reveal only small clusters of ethnic Chinese present in the area, primarily in the back streets, in local housing; at other times of the week it shifts to becoming a predominantly Chinese district; with a peak presence during the Sunday lunch hour, when London's Chinese community converges on the area. (Of course, to refer to a 'Chinese community', when this population in fact encompasses people from myriad ethnic Chinese backgrounds, is itself mistaken). Clearly, in a context such as this, in which no single group predominates, shifts in spatial, temporal and ethnic mix will change the spatial pattern quite subtly over the course of time.[73]

Jonathan Raban has written about how in London urban neighbourhoods known as 'Italian' or 'Jewish' have no clear boundaries, nor are they inhabited exclusively by people of those backgrounds. This is an essential aspect of successful urbanity:

> They are more or less arbitrary patches of city space on which several communities are in a constant state of collision. A colourful and closely-knit minority can give an area its "character", while its real life lies in the rub of subtle conflicts between all sorts of groups of different people, many of whom are visible only to the denizen.[74]

In fact, a strong cultural community can exist independently of any geographical boundaries. As we saw above, London's Chinatown mostly functions as a locus for a transpatial community, but periodically the area takes on clear ethnic spatial boundaries. The same is even more the case with dispersed minority religious communities, who come together once a week for prayers.[75] While London likes to consider itself as a city of villages, the reality is that even though it may bear the spatial signature of its village past, from a cultural point of view it is anything but a set of atomised communities.

Maps of nationality and race refute the simplistic notion that segregation is bad, integration is good. The fact is that in many cities immigrants and minorities choose to live in localised clusters, yet at the same time maintain a variety of social ties outside of their immediate neighbourhood. There are critical differences between voluntary segregation, such as that typified by contemporary European cities, and the involuntary segregation that has taken place elsewhere in the world.

Notes

1. M. Banton, *The Coloured Quarter: Negro Immigrants in an English City* (London: Cape, 1955), p. 94.
2. G. Simmel, 'The Metropolis and Mental Life,' in *The Sociology of Georg Simmel: Translated, Edited and with an Introduction by Kurt H. Wolff*, ed. KH Wolff (New York and London: Macmillan Publishing, 1950), p. 410.
3. L. Wirth, *The Ghetto (1988 Edition, with a New Introduction by Hasia R. Diner)*, ed. R.H. Bayer, Studies in Ethnicity (New Brunswick and London: Transaction Publishers, 1988; first published Chicago: University of Chicago Press, 1928). Wirth's book considers the Jewish community of Chicago in contrast with its historical roots in the Frankfurt Ghetto, illustrating how the community in Europe was shaped by its social isolation, leading to the intensification of its distinctive community life and culture. Wirth predicted that so long as hostility continued in the USA, some of those distinctive elements would endure.
4. L. Wirth, 'Urbanism as a Way of Life,' *The American Journal of Sociology* 44, no. 1 (1938), p. 12.
5. See e.g. L. Wacquant, 'A Janus-Faced Institution of Ethnoracial Closure: A Sociological Specification of the Ghetto,' in *The Ghetto: Contemporary Global Issues and Controversies*, ed. R. Hutchison and B. Haynes (Boulder, CO: Westview Press, 2011).
6. See M. Duneier, *Ghetto: The Invention of a Place, the History of an Idea* (New York: Farrar, Straus & Giroux, 2016), p. 22. Not to mention the cynical use of the term by the Nazis to describe their systematic segregation of Jewish populations under their conquest. The Nazi ghetto was not just a separating device; it constituted one of the first steps towards genocide.
7. C. Peach, 'Slippery Segregation: Discovering or Manufacturing Ghettos?,' *Journal of Ethnic and Migration Studies* 35, no. 9 (2009), p. 1388. Peach's writing on this subject is vital for understanding these distinctions between types of segregation.
8. Hillier, 'Cities as Movement Economies.' See also discussion on this feature of nineteenth-century London in Chapter 3.
9. This is seen in an historical analysis of Rome, which found that the city street could be 'regarded as a space of accord, both as a metaphor and as the place where such accord is practised daily' due to the fact that Roman citizens came from anywhere and everywhere. R. Laurence and D.J. Newsome, *Rome, Ostia, Pompeii: Movement and Space* (Oxford: Oxford University Press, 2011), p. 41.
10. N. Shah, *Contagious Divides: Epidemics and Race in San Francisco's Chinatown* (Berkeley: University of California Press, 2001), p. 17.
11. Shah, *Contagious Divides*, p. 29.
12. Farwell, W.B. *The Chinese at Home and Abroad Together with the Report of the Special Committee of the Board of Supervisors of San Francisco on the Condition of the Chinese Quarter of That City*. San Francisco: A. L. Bancroft & Co., 1885.
13. The map is stylistically similar to the Sanborn Fire Insurance maps of that period in the US and the Goad Fire Insurance plans in the UK, where buildings are coloured according to the degree of fire hazard of the building material and its content. Conveniently for this map's portrayal of the moral panic associated with the [sic] 'yellow peril' anti-Chinese propaganda, fire insurance plans would typically have the least hazardous material (brick) in pink and the most fire risky material (wood) in yellow.
14. The appendix lists 827 Dupont Street, for example, as follows: 'First storey, front, rear of Chung Wing & Co.'s dry goods store. One heavy iron door, rear first story; entrance through two iron doors with 3-inch plank; entrance from street and store, kitchen from rear by stairs

to second storey, about 16'x16' through heavy iron trap door; iron partition between store and gambling-room.' Farwell, *The Chinese at Home and Abroad*, p. 86.
15. Farwell, *The Chinese at Home and Abroad*, p. 39.
16. Farwell, *The Chinese at Home and Abroad*. The David Rumsey Historical Map Collection notes on the latter version of the map state that it was most likely printed for the use of key city officials. See more information on the map versions: http://www.davidrumsey.com/luna/servlet/detail/RUMSEY~8~1~241649~5512689:Official-Map-of-Chinatown-in-San-Fr. Accessed 24 July 2017.
17. The report's conclusions actually emphasise the need to enforce Californian laws (such as fire ordinances and labour laws) even though these are supposedly anathema to the Chinese population: 'the more rigidly this enforcement is insisted upon and carried out the less endurable will existence be to them here, the less attractive will life be to them in California. Fewer will come and fewer will remain . . . Scatter them by such a policy as this to other States . . .' (Farwell, *The Chinese at Home and Abroad*, pp. 67–8).
18. Farwell, *The Chinese at Home and Abroad*, pp. 39–40.
19. These were mainly Calvinist Protestants, fleeing religious persecution from Catholic France. B. Cottret, *The Huguenots in England: Immigration and Settlement 1550–1700* (Cambridge: Cambridge University Press, 1991).
20. A. Kershen and L. Vaughan, 'There Was a Priest, a Rabbi and an Imam . . . : An Analysis of Urban Space and Religious Practice in London's East End, 1685–2010,' *Material Religion* 9, no. 1 (2013). See also the comparative analysis of synagogues and churches in London's East End in the section 'Out of the Ghetto', p. 144.
21. Indeed, at the time of the study the area of high-density Jewish settlement was given a chapter of its own, 'The Chicago Ghetto,' which sat alongside chapters on 'The Bohemian People in Chicago' and 'Remarks upon the Italian Colony in Chicago'.
22. Note that this population also spilled into the map's survey area itself. Residents of Hull-House, *Hull-House Maps and Papers*.
23. Residents of Hull-House – a Social Settlement, *Hull-House Maps and Papers*, pp. 60–1.
24. L. Vaughan and A. Penn, 'Jewish Immigrant Settlement Patterns in Manchester and Leeds 1881,' *Urban Studies* 43, no. 3 (March 2006), p. 660.
25. P. Mateos, 'Uncertain Segregation: The Challenge of Defining and Measuring Ethnicity in Segregation Studies,' *Built Environment* 37, no. 2 (2011).
26. R.A. Woods, ed. *The City Wilderness: A Settlement Study by Residents and Associates of the South End House Edited by Robert A. Woods, Head of the House, South End Boston* (Boston and New York: Houghton, Mifflin and Company, 1899), p. v. The maps can be viewed at Harvard University Library online at http://ocp.hul.harvard.edu/dl/ww/HUAM57824soc.
27. The guide aimed principally to assist in finding the best charitable institution for any individual case of need, but was also intended to assess any gaps in charitable provision in the city. Bureau of Associated Charities, N.J, and A.W. MacDougall, *The Resources for Social Service, Charitable, Civic, Educational, Religious, of Newark, New Jersey; a Classified and Descriptive Directory* (New York: G. P. Putnam's Sons, 1912).
28. Another curiosity that is also worth mentioning is the Community Settlement Map, drawn up in 1976 for the city of Chicago. It is likely that it was drawn up by the then mayor, Richard Daley, as part of the city's bicentenary celebrations. The map uses free-form coloured shapes to record 23 communities (defined quite idiosyncratically, so 'Jewish' and 'Black' sit alongside ethnic or nationality-defined groups). It can be viewed at https://commons.wikimedia.org/wiki/File:Chicago_Demographics_in_1950_Map.jpg.
29. The Pale of Settlement was a region of imperial Russia which was the only area in which Jewish people were permitted to reside permanently. Dependent on charity and lacking the status of citizens, their move westwards was in part a desire to find a place to belong. See further information on life in the Pale and on the 1881–4 pogroms in J.D. Klier, 'What Exactly Was a Shtetl?,' in *The Shtetl: Image and Reality*, ed. G Estraikh and M. Krutikov (Oxford: Legenda, published by the European Research Centre, 2000), pp. 32–3; Kershen, *Uniting the Tailors*, pp. 9–10.
30. The book was published by the Toynbee Trust, which was founded at the London Settlement House of Toynbee Hall after the death of Arnold Toynbee, as a memorial to his work in promoting the investigation and diffusion of political and social economy.
31. Council of the United Synagogue, East End Scheme, June 1898, p. 39, cited in J. White, *Rothschild Buildings: Life in an East End Tenement Block 1887–1920* (London: Pimlico, 2003), p. 61.

32. C. Booth, *Life and Labour of the People in London*, 3rd series, 17 vols. (London: Macmillan and Co, 1903), vol. 4, p. 46.
33. Anonymous, 'Jewish East London,' *The Jewish Chronicle*, 26 April 1889.
34. See C. Russell and H.S. Lewis, *The Jew in London (with a Map Specially Made for This Volume by Geo. E. Arkell)* (London: Fisher Unwin, 1901). (Notes on the Map), pp. xxxiii–xlv.
35. Russell and Lewis, *The Jew in London* (Notes on the Map), p. xl.
36. W.J. Fishman, *East End 1888: A Year in a London Borough among the Labouring Poor* (London: Gerald Duckworth & Co. Ltd, 1988), p. 11.
37. This is borne out statistically, with a bifurcation between the streets where immigrants were a minority (up to 50 per cent, namely the streets coloured red), which become more accessible (integrated, in space syntax terminology) the denser they become, and the streets where the immigrants were a majority (the streets coloured blue), which were less accessible as Jewish density increased. See Chapter 3 for an illustration of the space syntax analysis of the 1890s map. Full analysis of the map of Jewish East London can be found in Vaughan, '*The Relationship between Physical Segregation and Social Marginalisation in the Urban Environment*.'
38. Booth, 'Poverty Series Survey Notebooks (Online Archive)', BOOTH/B/B356, pp. 122–3.
39. Booth, 'Poverty Series Survey Notebooks (Online Archive)', BOOTH/B/B351 pp. 100–1.
40. Jewish employers would occasionally provide charitable and other financial support to their poorer neighbours and employees; see Davin, *Growing up Poor*; White, *Rothschild Buildings*.
41. B. Walter, 'Irish/Jewish Diasporic Intersections in the East End of London: Paradoxes and Shared Locations,' in *La Place De L'autre*, ed. M. Prum (Paris: L'Harmattan Press 2010), p. 59.
42. D. Englander, *A Documentary History of Jewish Immigrants in Britain 1840–1920* (Leicester: Leicester University Press, 1994), p. 64. See also: 'To those who lived in Flower and Dean Street, Bethnal Green was merely at the top of Brick Lane . . . "we avoided it because we were afraid of being beaten up. . ."' White, *Rothschild Buildings*, p. 136.
43. R. Johnston, J. Forrest and M. Poulsen, 'Are There Ethnic Enclaves/Ghettos in English Cities?,' *Urban Studies* 39, no. 4 (2002), p. 609.
44. Russell and Lewis, *The Jew in London*, p. 19.
45. The analysis of the places of worship was published in L. Vaughan and K. Sailer, 'The Metropolitan Rhythm of Street Life: A Socio-Spatial Analysis of Synagogues and Churches in Nineteenth Century Whitechapel,' in *An East End Legacy. Essays in Memory of William J Fishman*, ed. C. Holmes and A. Kershen (London: Routledge, 2017).
46. The study found that putting aside those that were completely hidden, the study area's 14 synagogues were characterised by a significantly more constrained viewshed (isovist) from their front entrance (28–133 metres) than the 7 churches in the study area (139–1093 metres).
47. Nightingale, *Segregation*, p. 93.
48. B. Tecle-Misghina, *Asmara – an Urban History: Rivista L'architettura Delle Città – Unesco Chair Series N. 1* (Rome: Edizioni Nuova Cultura, 2015). The map we see illustrated here, from 10 April 1916 (scale around 1:5000) was drawn by O. Cavagnari for the Government of the Colony of Eritrea, printed in Rome for the Ministry of the Colonies, General Directorate for Political Affairs, Map Office.
49. Tecle-Misghina, *Asmara*.
50. Tecle-Misghina, *Asmara*, p. 50.
51. E. Denison, M. Teklemariam and D. Abraha, 'Asmara: Africa's Modernist City (Unesco World Heritage Nomination),' *The Journal of Architecture* 22, no. 1 (2017), p. 19.
52. Denison, Teklemariam, and Abraha, 'Asmara', p. 32.
53. W.E.B. Du Bois, *The Souls of Black Folk* (Chicago: McClurg & Co, 1903), eBook version. Chapter 9, *Of the Sons of Master and Man*.
54. Du Bois, *The Souls of Black Folk*. Chapter 9.
55. 'Report no. 9, Summary, Re-survey of Denver, Colorado by the Division of Research & Statistics,' via Denver Public Library webpage https://history.denverlibrary.org/news/new-whg-redlining-maps-denver. Accessed 17 July 2017.
56. P. Bennett, 'Geographies of Financial Risk and Exclusion,' in *The Sage Handbook of Social Geographies*, ed. S.J. Smith et al. (SAGE Publications, 2009), pp. 228–9.
57. A. Hillier, 'Searching for Red Lines: Spatial Analysis of Lending Patterns in Philadelphia, 1940–1960,' *Pennsylvania History: A Journal of Mid-Atlantic Studies* 72, no. 1 (2005). It should be pointed out that Hillier's research suggests that redlining didn't *cause* discrimination, but instead reflected the discriminatory practices prevalent at the time.

58. Hillier, *Searching for Red Lines*, p. 48.
59. Bennett, 'Geographies of Financial Risk and Exclusion,' p. 229.
60. K.-M. Cutler, 'East of Palo Alto's Eden: Race and the Formation of Silicon Valley,' *TechCrunch.com* (2015), http://techcrunch.com/2015/01/10/east-of-palo-altos-eden/. See also the work of the National Community Reinvestment Coalition, which overlays contemporary income and minority population spatial statistics over the historic valuation maps, finding a considerable spatial continuity between redlining and poverty and race: http://maps.ncrc.org/holc/.
61. A. Lemon and D. Clifford, 'Post-Apartheid Transition in a Small South African Town: Interracial Property Transfer in Margate, Kwazulu-Natal,' *Urban Studies* 42, no. 1 (2005). See also C. Spinks, 'A New Apartheid? Urban Spatiality, (Fear of) Crime, and Segregation in Cape Town, South Africa'. Development Studies Institute Working Paper Series No. 01-20 (London: London School of Economics, 2001).
62. R.J. Sampson, *Great American City: Chicago and the Enduring Neighborhood Effect* (Chicago: University of Chicago Press, 2012).
63. H. Rittel and M. Webber, 'Dilemmas in a General Theory of Planning,' *Policy Sciences* 4, no. 2 (1973).
64. R. Sennett, *The Uses of Disorder: Personal Identity and City Life* (London: Faber and Faber, 1996).
65. The Jewish quarter was called a 'ghetto' both by its inhabitants and by outsiders. Most famous among the former was perhaps Israel Zangwill, whose novels on life in the East End at the turn of the twentieth century open a door on its interior world. See for example: 'This synagogue was all of luxury many of its Sons could boast. It was their salon and their lecture-hall. It supplied them not only with their religion but their art and letters, their politics and their public amusements. It was their home as well as the Almighty's. . . It was a place in which they could sit in their slippers, metaphorically that is; for though they frequently did so literally.' I. Zangwill, *Children of the Ghetto: A Study of a Peculiar People* (London: Heinemann, 1922, first published 1892), pp. 141–2.
66. Booth, *Life and Labour of the People of London*, 1: East, Central and South London, pp. 66–7.
67. B. Walter, 'England People Very Nice: Multi-Generational Irish Identities in the Multi-Cultural East End,' *Socialist History Journal* 45 (2014).
68. Dennis, *Cities in Modernity*, p. 66.
69. J. Freedman, *Immigration and Insecurity in France* (S.l.: Taylor & Francis, 2017), p. 155; C.-U. Schierup and A. Ålund, 'The End of Swedish Exceptionalism? Citizenship, Neoliberalism and the Politics of Exclusion', *Race & Class* 53 (2011), 45–64.
70. A. Kershen, 'The Construction of Home in a Spitalfields Landscape,' *Landscape Research* 29, no. 3 (2004), p. 265.
71. C. Roden, 'Food in London: The Post Colonial City' (paper presented at 'London: Post-Colonial City,' a meeting of the Architectural Association, London, 12–13 March 1999).
72. In fact, it is very rare to have both recorded. An exception to this is Bill Williams's study of workplaces in Manchester's nineteenth-century Jewish quarter. See B. Williams, *The Making of Manchester Jewry 1740–1875* (Manchester: Manchester University Press, 1985). See also L. Vaughan, 'The Unplanned "Ghetto": Immigrant Work Patterns in 19th Century Manchester' (paper presented at 'Cities of Tomorrow', the 10th conference of the International Planning History Society, Westminster University, July 2002).
73. S. Chung, 'London Chinatown: An Urban Artifice or Authentic Chinese Enclave?' (paper presented at the 1st City Street International Conference, Notre Dame University, Louaize, Lebanon, 18–20 November 2009).
74. J. Raban, *Soft City* (Glasgow: William Collins & Sons, 1974), p. 184.
75. C. Dwyer, D. Gilbert and B. Shah, 'Faith and Suburbia: Secularisation, Modernity and the Changing Geographies of Religion in London's Suburbs,' *Transactions of the Institute of British Geographers* 38, no. 3 (2013).

6
Crime and disorder

A thoroughly vicious quarter. The presence of the Cambridge Music Hall in Commercial St. makes it a focussing point for prostitutes . . . North up Wilkes St. . . . is a lodging house frequented by ex-convicts. "Six or seven old 'lags' living there now". No shame about having been in prison here, one came across street to check whether his license had come yet.[1]

Moral geography

The previous chapter showed how easily the terms 'ghetto' and 'slum' are conflated such that the negative attributes of a number of streets might be assumed to apply to entire areas, and thus to everyone living within them. In this chapter we will see how instances of disorder, deviant behaviour and indeed crime are collectively used to label areas as problematic, so as to cause an ecological fallacy to take hold. The subtler maps will be more cautious in doing so, but still, even today, we see the problematic use of crime maps to obfuscate the specificity of crime: areas are written off as prone to crime, without getting to grips with the underlying causes of crime nor the nature of that crime.

In the early part of the nineteenth century, techniques in mapping statistics started to allow scientists to pinpoint the location of clusters of urban problems. Innovations in mapping in the latter part of the nineteenth century brought to the attention of the public the scale of the problem, as well as how it might be ameliorated.

From Mayhew and Booth through to the criminological practice known as 'broken windows' today (that argues that broken windows are signifiers of a problem area),[2] the social ranking of places by signs of physical disorder has a long history:

> At the west end of Dorset St. leading into Brushfield St. is Little Paternoster Row. Black on both sides in map on East side only. 2 & 3 storied common lodging houses. Ragged women & children. Holey toeless boots. windows dirty, patched with brown paper & broken. Prostitutes, thieves & ponces.[3]

While one might argue with the criticism of Booth's maps of poverty as being mere topographies of morality, it is fair to say that by displaying how vice, crime, disorder and so on were arranged spatially these maps did highlight areas as problematic and hence present the clearance or tidying up of slum areas as a solution.[4]

Felix Driver has written about how 'the watchword of nineteenth-century social science was "improvement"; social science became a form of philanthropy, whose chief strategy was to find the means to ameliorate the living conditions of the poor'.[5] In fact, as Driver points out, though many writers on the history of the social sciences make a distinction between the two, it is quite difficult to distinguish between the activities of nineteenth-century scientists and those of the reformists. Especially in the case of the 'environmentalists', namely the scientists who mapped statistics onto maps, mapping immoral behaviour onto the urban environment, they were in effect blaming the dirty or diseased environment for the 'vicious' or ill health of the people it contained. By ameliorating the environment, its inhabitants could supposedly be improved in one fell swoop.

The success of medical topography was highly influential in this sense (as we saw in Chapter 2), given that by mapping patterns of disease it was possible to arrive at correlations between ill health and aspects of climate, topography or drainage. Once medical geographers moved into cities and became sanitary scientists, it was an easy shift to move from the reform of housing conditions (and their associated diseases) to intervention into social conditions (and their associated poor morals). Driver cites a quote from the London Statistical Society in 1849 that recommends improving factors such as education and policing to create a healthy *social* fabric.[6]

The prevailing concern was that the dense, dark urban fabric would conceal the worst of the social indecencies of the rookeries, whose interiors, as Robin Evans has pointed out, were 'characteristically portrayed as the scene of daylight dissipation, drunkenness and criminal conspiracy . . . a picture not of an actual place but of a latent condition'.[7] The association between physical and moral degradation went deeper still; it was the immoral habits of the poor which were said to fester in (and be fostered by) the worst of the tenement dwellings: 'filthy habits of life were never far from moral filthiness'.[8] Not only that, but the potentially contagious nature of the immorality to be found in the most physically segregated corners of the city, provided places where the moral disease was located beyond the gaze of the public and beyond the control of the police. The conviction was that 'virtue could be wrought from architecture as surely as corruption was wrought from slums'.[9] The physical solutions to these festering sores on the body of the city were termed 'bridges' by the editor of the journal *The Builder*. They were to be scientifically based solutions that, by the mid-nineteenth century, centred on constructing institutions such as model dwellings and Ragged Schools as well as public parks. At the same time, the conviction was that by opening up the urban 'labyrinths' of courts and cul-de-sacs the new urban order would expose the secret haunts of the immoral poor to the light of day in order to make them more amenable to moral improvement.[10]

As in London, elsewhere in other rapidly growing cities similar reforms were in train. Hell's Kitchen (a midtown district of Manhattan, New York) had acquired a notorious reputation by the end of the nineteenth century. Joseph Varga describes how various police practices, backed up by statutes relating to 'disorderly conduct', were applied in specifically 'frozen zones', where criminality was unofficially left unpoliced. Over time, these areas then became the target for improvement of their physical conditions.[11] The sort of tenement houses we have already seen, designed by the New York city legislators to cleanse the area of disease, were meant at the same time to achieve concomitant moral and social improvements. Joseph Varga argues that this had only partial success, stating that the externally constituted boundary of the new communities transmuted over time to become a demarcation of an area that contained negative behaviour, such as gang activity.[12] Evidently once it was conceded that there might be an association between environmental degradation and social vice it was a short step to find social vice where there were signs of environmental degradation – particularly dirt. Accessibility then, 'through movement', was viewed as 'good' in normative terms by middle-class urban elites, a way of cleaning up the city and making it more governable.

By the time of the Chicago School, this approach to ameliorating urban crime and deviance had become fixed into a solution to the problem of slums (and, later on, 'ghettos'). Although proponents of this school recognised the ecological fallacy of associating individuals with the perceived 'derelict and vicious' character of their area, the tendency to see the evolution of cities as a natural process of sorting between advantageous and less advantageous locations, instead of a political outcome of racist or economic processes, started to take hold.[13] The influence of ideas of moral geography over the course of the last century, which transformed into notions of social disorder, are still extant to a certain extent today, when the regeneration of social housing is often justified by using terms such as 'blight' and when depictions of an area's dirt, graffiti and deterioration are sometimes used to justify wholesale destruction, rather than piecemeal regeneration.[14]

Contagion and morality

Although this book focuses on the past two centuries, it is important to bear in mind the position of prostitution as an aspect of urban crime and deviance that dates back centuries, if not millennia.[15] Bronislaw Geremek's history of *The Margins of Society in Late Medieval Paris* shows, for example, how prostitutes had designated zones, or entire streets on the edge of the city.[16] He writes that 'knowing where prostitution was practiced contributes important information about the social topography of the town', given that it was typically assigned to specific areas inside or outside of the walls.[17] This was a form of social hygiene that placed vice in close proximity to poverty and well away from the more prosperous areas. Equivalent statutes applied to London, Venice, Dijon and other cities throughout Western Europe.

A slightly different sort of social hygiene meant that well into the eighteenth century, Portsoken, the easternmost ward in the City of London and the only one completely outside of its walls, became the preferred place of settlement for incomers, such as French Huguenots or Jewish merchants. The ward was as close as these new arrivals could get to the City's regulated markets without actually living inside it (see Figure 6.1, map of Portsoken, 1772).[18]

Gilles Palsky has recounted how, from their invention in the early decades of the nineteenth century, most thematic maps related to the natural sciences. Yet with the emergence of human sciences for the study

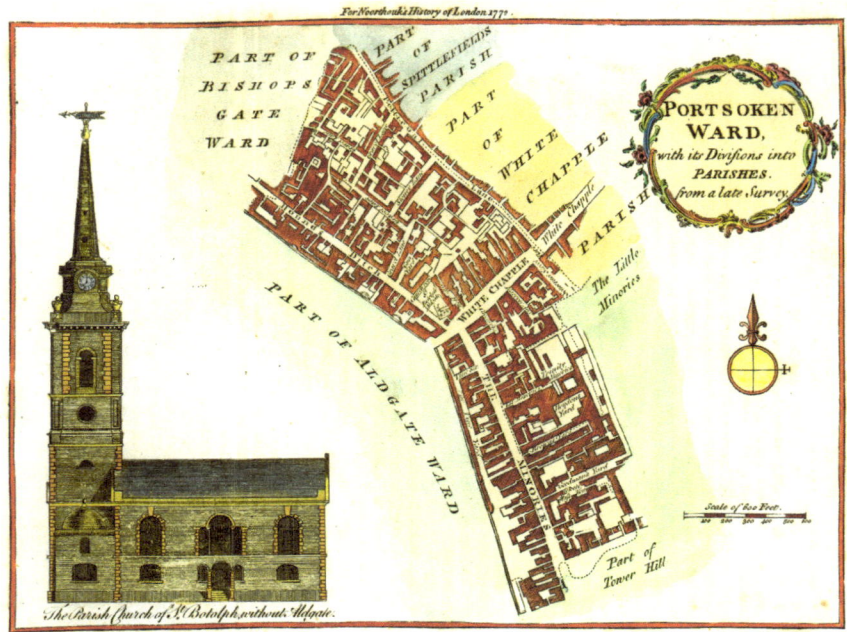

Figure 6.1 *Portsoken Ward with its Divisions into Parishes. From a Late Survey.*
Published in J. Noorthouck, *History of London* (London: privately printed, 1772). Image copyright Jonathan Potter Ltd.

of human society, the rapid development of graphic innovation launched a century of developments in social cartography, much of which has been described in this book's preceding chapters.[19] The many maps which were drawn up throughout the century allowed for the spatial patterns of social statistics to be described and summarised, and even allowed the discovery of new facts: they enabled differences to be made visible.

Pierre Charles Dupin's invention of the choropleth map in 1826, which presented an analysis of the distribution and intensity of illiteracy in France in shaded areas from black to white, was possibly the first modern statistical map.[20] Three years later, André Michel Guerry created the first comparative choropleth thematic maps, showing crimes against persons and crimes against property in relation to level of education, across the departments of France. His method took hold first in France and then spread to English-speaking countries with the translation of his work from the French.

This was a period during which France led the world in capturing social statistics on maps. One example, discussed by Palsky, *Distribution des*

Prostituées dans chacun des 48 quartiers de la Ville de Paris [Distribution of Prostitutes in each of the 48 quarters of Paris, 1836], captured an ostensibly neutral picture of the spatial distribution of data on the relative density of prostitute populations across Paris that had been provided by its police force. These were not, as Palsky has shown, 'neutral illustrations, but were primarily conceived as arguments in scientific or ideological debates, and . . . their sign system played a major role in their persuasive effect'.[21]

The maps featured in the book *De la Prostitution dans la ville de Paris considérée sous le rapport de l'hygiène publique, de la morale et de l'administration* [Prostitution in the city of Paris: considered in terms of public hygiene, morals and administration] by Alexandre Parent-Duchâtelet.[22] Its telling of the history of prostitution in the city since the sixteenth century was highly influential and its maps were innovative in capturing data graphically; indeed they provided a visual underpinning to the interpretation of the moral statistics captured in the book.[23] Yet Duchâtelet's research was influential in its own right – despite a use of physical stereotyping that is jarring to the modern reader (he attributes plumpness of figure to laziness and greed, a raucous voice to social origin, abuse of alcohol and exposure to cold) – especially his radical thinking on the social causes of prostitution. His conviction that the main reasons for turning to prostitution were related to poverty and that poverty was a function of the social conditions of the time show an unusual sympathy towards the human condition of those on the margins of society.

Reading Michael Ryan's report from 1839 on the comparative merits of how prostitution was managed in France, America and Britain, it is interesting to see how he praises Duchâtelet's book for its detail on how Paris managed to control the vice of prostitution through spatial means. Paris was at the forefront in regulating prostitution spatially. Ryan details how in Paris brothels were not allowed near sacred buildings, palaces, residences of high functionaries, schools or hotels.[24] Nor were brothels permitted within the visual ambit of schools, unless they were beyond a bend in the street; it was preferred that they be situated within narrow, 'little frequented', thinly populated streets. Moreover, in 'streets or courts which end in a wall, or have no thoroughfare, and which no one would attempt to enter with a view of abridging his road, the police never refuse to tolerate them, when asked'.[25]

The fear of moral (let alone physical) contagion is a constant refrain throughout Ryan's book, which describes the risk that the 'daughters of the ignorant, depraved, and vicious part of our population' might 'enter upon a life of prostitution for the gratification of their unbridled passions, and become harlots altogether by choice. These have a short career, generally dying of the effects of intemperance and pollution, soon after entering upon this road to ruin'.[26] This is a fear not dissimilar to that of those who wrote about San Francisco's Chinatown, where the reader is not always entirely clear whether the fear of contagion is just as much about immorality as about the physical aspects of disease:

> The lowest form of prostitution – partaking of both slavery and prostitution – they have planted and fostered to a lusty growth among us, and have inoculated our youth not only with the virus of immorality in its most hideous form but have, through the same sources, physically poisoned the blood of thousands by the inoculation with diseases the most frightful that flesh is heir to.[27]

Later in the century we can see how for Booth's policemen, interviewed in the 1898–9 survey notebooks, prostitution was a common feature of the worst of his streets. In fact, the descriptor of Booth's lowest rank of poverty centred on the concept: 'Black. The lowest grade (corresponding to Class A), inhabited principally by occasional labourers, loafers, and semi-criminals – the elements of disorder'. Spatial disorder was related to social disorder, dirt to immorality. Note the many references to courts, or a narrowing of a street, in the following passage:

> Into Dorset St. black in map. still black . . . thieves, prostitutes, bullies. All come from lodging houses. Some called "doubles" with double beds for married couples, but merely another name for brothels: women, draggled, torn skirts, dirty, unkempt, . . . Jews standing about in street or on doorsteps . . . The back rooms of a common lodging house for men & women come through into this ct. from the front of the street, open drains & taps on the west side of the Ct. for washing purposes. The next East is Old Dorset Court. Has been done up.[28]

> North & East along Old Montague Street. Narrow, long, 3 storied. Houses. Shops. Owners living above. Pink on map purple. Many courts out of the South side. All Jews. Going Eastwards. Green Dragon Place. Poor Jews, used to be rough. Now lb. rather than db. of map, no turning on the East side. Montague Place. 4 houses only, 2 st. homework, cement paved, clean, lb to purple, d blue in map. Easington Bldgs. 15, 2 storied houses. Narrower & poorer looking.

> Dirty children, lb. db. in map. Kings Arms Court. Db in map, now no dwelling houses, great mess, old boots, tins, orange peel, onion skins, paper. Black Lion Yard still dark blue as map, mixture of Irish & Jews. Rough; houses 4, 3 & 2 storied, some small shops.[29]

As we saw in the analysis of the Booth maps in Chapter 3, Booth and his team's reading of spatial conditions stemmed from an intuition that has subsequently been borne out with spatial analysis. Their frequent allusions to the problematic layout of poverty areas was an expression of their concerns regarding the characteristic of many areas of deepest poverty: that they were spatially segregated, cut off from the lifeblood of the city. Booth's professed desire to tidy up the streets, to bring spatial order to the social disorder, stemmed from his correlating of complicated spatial arrangements with the presence of disorderly conduct. Even though this spatial determinism was essentially incorrect, in the sense that the layout did not *directly* cause the social outcome, the underpinning idea, that social inequity was shaped or supported by a lack of spatial connectivity, has a significant basis in fact.

Almost contemporaneous with Booth's maps, and indeed with the Hull-House project in Chicago, was the work of William Thomas Stead in the latter city. Stead was probably Britain's first investigative journalist;[30] he had established by this time a record for campaigning newspaper articles against poverty and prostitution, including promoting in 1883 the polemical pamphlet released by the Rev. Andrew Mearns, 'The Bitter Cry of Outcast London', through the pages of the *Pall Mall Gazette*. Stead's journalism was highly successful, contributing to the setting up of a Royal Commission on the Housing of the Poor in 1884. In contrast with the measured tones of the Booth and Hull-House projects, Stead's text is distinctive in its use of sensationalist language, with lurid descriptions of the disreputable and the degenerate occupying the worst districts of the city of Chicago.[31] Having travelled to the United States to attend the World's Fair of 1893 in the city, Stead published the 500-page 'If Christ came to Chicago! A Plea for the Union of All Who Love in the Service of All Who Suffer', in which he proclaimed that faith in Christ by every town-dweller would 'lead directly to the civic and social regeneration of Chicago or any other great city'. The map drawn up for Stead's book (Figure 6.2) depicted the worst of the precincts he studied. Not only was the choice of precincts intentional, so were the boundaries he selected for the map itself, which emphasises the dominance of prostitution within the precinct by choosing a tightly defined set of streets. By colouring brothels in red he drew the eye to the scale of the problem. The dry tones of the map

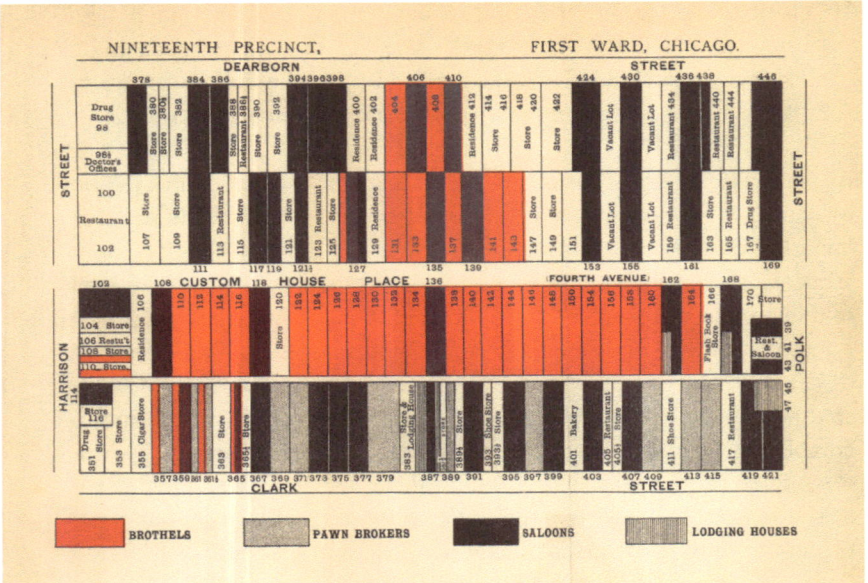

Figure 6.2 *Nineteenth Precinct, First Ward, Chicago*, 1893.

From W.T. Stead, *If Christ Came to Chicago – a Plea for the Union of All Who Love in the Service of All Who Suffer* (London: The Review of Reviews, 1894). Copyright Cornell University – PJ Mode Collection of Persuasive Cartography.

were not entirely guileless. Contemporary records of his extended stay in Chicago describe how, following his visits to Fourth Avenue (central on the map), effectively the heart of the city's red-light district, Stead stirred up controversy amongst the city's dignitaries by addressing meetings in its most respectable bastions, such as the Women's Club, where he accused its members of being more disreputable than a harlot because of the self-indulgence of their style of living, which ignored the poverty surrounding them.[32]

Stead was using his best weapons – shocking language and sensationalist imagery – to get across his frustration with the depths of vice and immorality that he had seen in the city. Similarly, the focus on the brothels in the pull-out map to his book was much more shocking at the time than can be comprehended today: he was in effect providing a directory of vice, a dramatic contrast with the Christian language suggested by the book's title.

Stead's description of his map of the nineteenth precinct, with its 'forty-six saloons, thirty-seven houses of ill-fame and eleven pawnbrokers', which he points out as the 'moral sore spots of the body politic', maintains that it in fact underestimates the reality of the problem. It gives, he wrote, 'an unduly favourable impression', as many of the stores and offices are 'more or less haunted by immoral women', while the precinct has so many saloons that it is impossible to not become intoxicated. Stead's writing becomes Wesleyan in his subsequent description of the area, where you 'look in vain' for 'any bath or washhouse where cleanliness, which is next to godliness, can be cultivated'.[33]

In many ways these views of prostitution have not changed for centuries. Medieval views of prostitution as being associated with defilement and disease meant that despite the relative toleration for prostitutes, their physical marginalisation placed them outside the structures of society. Nineteenth-century Paris and London similarly viewed the prostitutes themselves as the 'refuse and filth of society',[34] while in the contemporary United States, popular culture has red-light districts viewed – by the main protagonist in the film *Taxi Driver* – simply as 'filth'.

Contemporary regulations seem to tolerate prostitution as long as it is not out of place. In fact, the 'out of placeness' of prostitution as being akin to filth is, by a little stretch, akin to the anthropologist Mary Douglas' own definition of dirt as being 'matter out of place' – not in the simplistic sense of dirt being in opposition to cleanliness, but in the sense that disgust towards the disordered or the dirty is part of a need to make a separation between the clean and the defiled.[35] Prostitution is considered to be 'in-place', as Hubbard puts it, when it takes place in the economically (and physically) marginal spaces of the city. Whether in medieval times or today, removing the polluting, deviant behaviour of society so that it is out of sight seems to continue to be the acceptable way of treating prostitution in most western cities.[36] One gets the sense from writings such as those of Stead that the vehemence of his language and the strength of his moral geography, is due to his recognition that the only way to improve the problem of prostitution (and the more general arduous struggle of the district's denizens against poverty) is to drag urban problems out of the city's dark corners into public view so that people can no longer ignore them.

Drink: a modern plague?

We saw in Chapter 3 how the Reverend Abraham Hume drew up a series of four maps showing the 'Condition of Liverpool, Religious and Social' in a self-published booklet from 1858. Hume listed a large number of public houses in the borough of Liverpool in his booklet – one for every 307 individuals, with many districts having far more than their fair share.

He found a distinct geography of poverty, typically located in clusters a single turning off the main roads, yet crime was more tightly clustered still. He found criminals present in 33 of all the streets he studied, in contrast with poverty, which was present in 195 streets in total. These clusters 'specially devoted' to crime, vice and immorality were a focus for police concern, who reported that 'the professional prostitute is always the ally of thieves'.[37]

Hume was at pains to make a distinction between *crime*, which might lie beyond clerical influence (though a Christian rebuke might have some effect), and *immorality*, which had the potential to be improved. Hume linked 'intemperance' with various other 'social evils' relating to immorality – the one being often the 'parent' of the other. Intemperance, meaning a lack of moderation or restraint, was a term that was in common usage by the mid-nineteenth century, so readily available to be associated with the temperance movement, namely the movement against immoderate indulgence in intoxicating drink.[38]

Intemperance was an important social concern during the latter half of the nineteenth century in Britain. Many organisations started to focus their efforts on social change, rather than changing the habits of individuals. James Kneale writes how the social context of drink – the role of the drink trade and rituals of conviviality ('treating', or buying a round of drinks amongst friends or acquaintances) – became common discussion points in temperance documents.[39] Once drinking became seen as a social problem and not a disease of the individual, it was a logical move to start to map the location of the problem in its urban context, which explains the proliferation of drink maps during this period, whose cartography represented alcohol 'as a *spatial* problem'.[40]

Charles Dickens pointed to another concern at this time: the replacement of small pubs with 'gin palaces'. A gin palace was essentially a large building that offered gin, instead of the traditional, small pub. His chapter on the subject in 'Sketches by Boz' opens with a droll description of the 'disease' of large gin shops being created by knocking several smaller ones together, attaining a 'fearful scale', with quiet old pubs being replaced by plate-glassed spacious premises:

> The gin-shops in and near Drury-Lane, Holborn, St. Giles's, Covent-garden, and Clare-market, are the handsomest in London. There is more of filth and squalid misery near those great thorough-fares than in any part of this mighty city . . . The hum of many voices issues from that splendid gin-shop which forms the commencement of

the two streets opposite; and the gay building with the fantastically ornamented parapet, the illuminated clock, the plate-glass windows surrounded by stucco rosettes, and its profusion of gas-lights in richly-gilt burners, is perfectly dazzling when contrasted with the darkness and dirt we have just left. The interior is even gayer than the exterior. A bar of French-polished mahogany, elegantly carved, extends the whole width of the place; and there are two side-aisles of great casks, painted green and gold, enclosed within a light brass rail . . .[41]

This is contrasted with the 'filthy and miserable appearance of this part of London' with 'wretched houses with broken windows . . . every room let out to a different family. . .' The gin palaces were, Dickens reported, full of 'drunken besotted men, and wretched broken-down miserable women'. In fact, he stated that though drunkenness was a great vice, poverty was a greater vice still:

> . . . and until you can cure it, or persuade a half-famished wretch not to seek relief in the temporary oblivion of his own misery, with the pittance which, divided among his family, would just furnish a morsel of bread for each, gin-shops will increase in number and splendour. If Temperance Societies could suggest an antidote against hunger and distress, or establish dispensaries for the gratuitous distribution of bottles of Lethe-water, gin-palaces would be numbered among the things that were [reduced in number and splendour]. Until then, their decrease may be despaired of.[42]

Temperance societies had a role to play, therefore, in mapping the 'disease' of gin palaces in a similar way to mapping yellow fever or cholera in a previous generation. An early example of this is the *Map Shewing the Number of Public Houses in the Metropolis*, which was presented in a paper about the connection between alcohol and crime read by John Taylor in 1860 'Before the National Association for the Promotion of Social Science' and compiled for the National Temperance League. Along with a red mark for each location selling alcohol in London, the map included data on the population of the city at the time of the last census (1851) as well as the number of dealers in alcohol.[43]

A more graphically effective pub map was compiled by the National Temperance League. It featured a map of a large area of London, covered in large pink spots, one for each pub (see Figure 6.3). Entitled *The Modern Plague of London*, it unabashedly presented London as if it were contaminated with a ghastly disease of the skin, with a pink pox covering

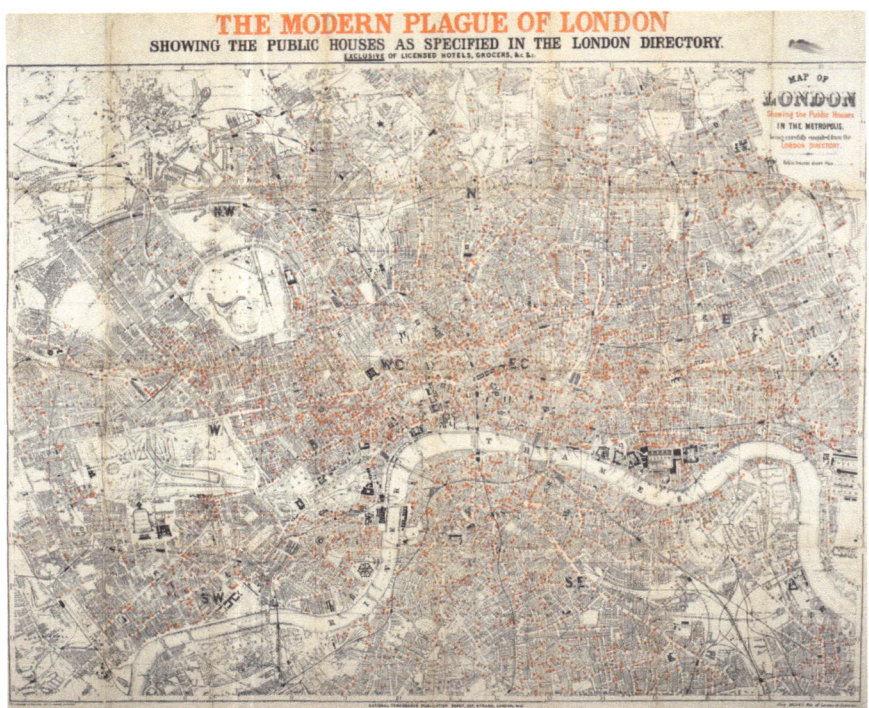

Figure 6.3 *The Modern Plague of London*, c.1884, showing the public houses as specified in the London Directory exclusive of licensed hotels, grocers & etc.

National Temperance League. Copyright Museum of London.

the central districts almost completely. The map's dots were drawn out of scale to emphasise the problem (see Figure 6.4, a detail of the above).

It is interesting to contrast this with Charles Booth's own more sober drink map (see Figure 6.5), which distinguished between on- and off-licences, charting the differences between landscapes of street-corner beerhouses in working-class neighbourhoods and the grander gin palaces on the main thoroughfares of the West End of London, and showing also the relationship between intemperance and an absence of places of worship. This map, the only stand-alone map included in Booth's vast study, was produced for the social science section of the Paris Exhibition of 1900:

> On it the five different forms of licensed premises are marked according to their character, excepting that in the case of the City [which had too many to mark at this scale] . . . to give to the ordinary reader, at a glance, an impression of the ubiquitous and manifold character of the three most important social influences.[44]

Figure 6.4 Detail of *Modern Plague of London*.
National Temperance League. Copyright Museum of London.

Two categories of elementary school are shown, along with the six categories of places of worship (five for different church denominations and one for synagogues) and five different types of 'Houses Licensed to Sell Intoxicating Drinks', depending on the nature of the license (see Figure 6.6). In many places the black circles of public houses completely outnumber the churches and schools. Reading through the volume on 'Notes on Social Influences' for which this map is an accompaniment, the commentary repeatedly makes this association, but also brings the social influences back to the physical conditions of the built environment: 'between evil conditions of health and evil conditions of life generally. The results of jerry building. More care exercised in paving and sweeping of streets and dust removal, but still much to be desired in poorer parts: refuse still dumped on the marshes'.[45]

A detail from the map covers an area of the West End (see Figure 6.7) which had a veritable epidemic of pubs; it is not a coincidence that this area was chosen for another map, by Joseph Rowntree and Arthur Sherwell for their highly successful 1899 book, *The Temperance Problem and Social Reform*.[46]

Brian Harrison has written about how the pub and the temperance society were in constant competition with each other for the attention of the growing masses of Victorian urban society. This was a new

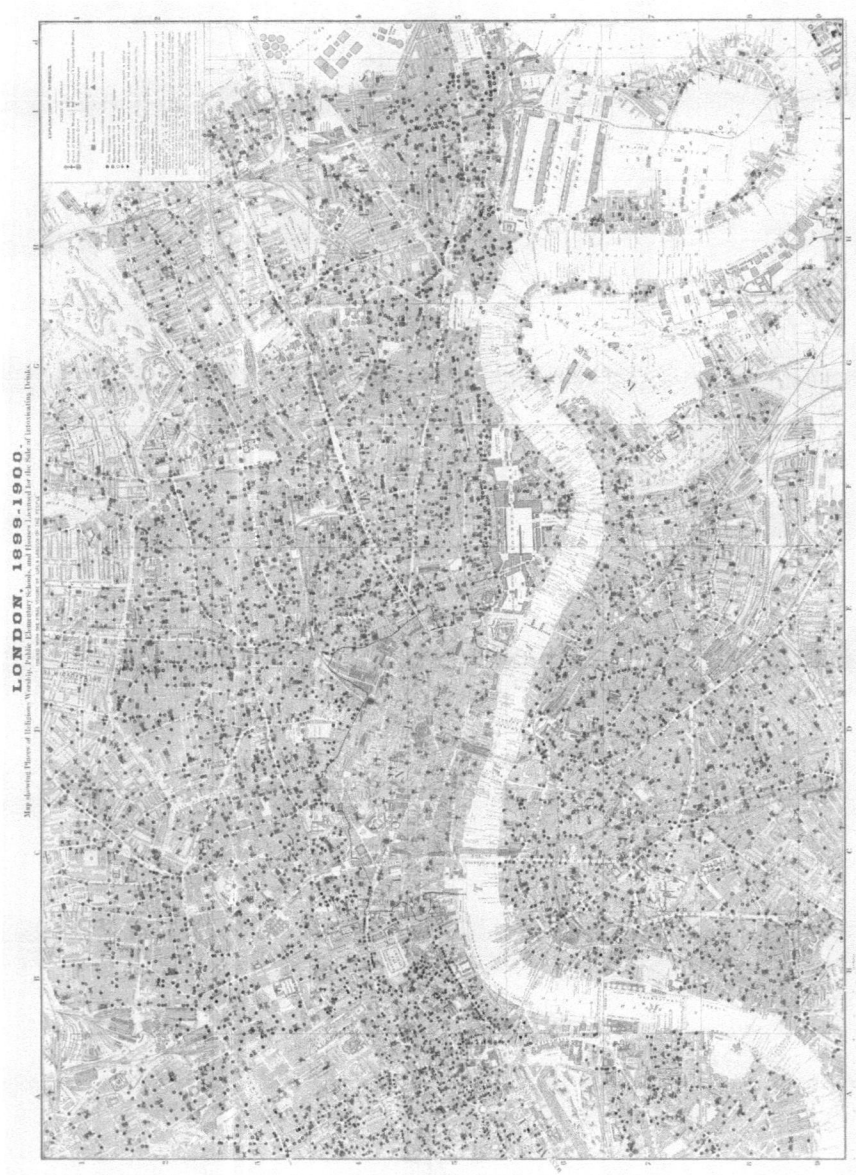

Figure 6.5 *Map Showing Places of Religious Worship, Public Elementary Schools, and Houses Licensed for the Sale of Intoxicating Drinks*, London, 1900.

Charles Booth, 1900. Copyright Cornell University – PJ Mode Collection of Persuasive Cartography.

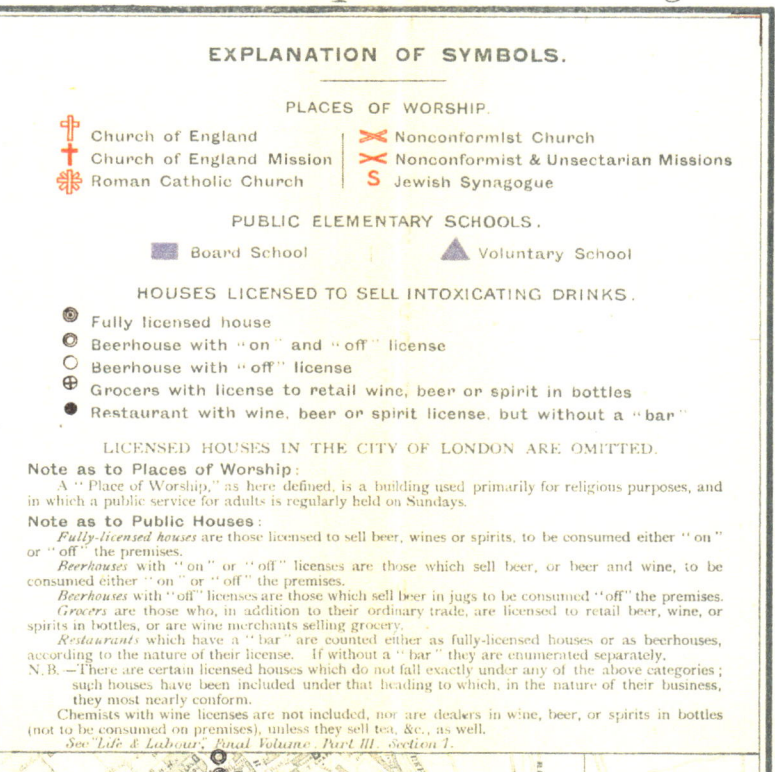

Figure 6.6 Key to *Map Showing Places of Religious Worship*, etc., London, 1900.

Figure 6.7 Detail of *Map Showing Places of Religious Worship*, etc., London, 1900.

Charles Booth, 1900. Copyright Cornell University – PJ Mode Collection of Persuasive Cartography.

development; in rural parts, there was a symbiotic relationship between pub and church, with church meetings taking place in pubs and churchgoers stopping off at the pub on the way to prayers. Parsons might even promote church-ales. Moving into the city the pub became central to social life, especially for the working-class man (as we saw to a lesser extent in Rowntree's work on York), who 'had to choose the life of the pub and the music-hall or the life of the temperance society, mutual improvement society, and chapel: there was nowhere else to go'.[47] Pubs were also stopping-off points for transportation and, prior to the building of the railways, coaching-inns took on the role of railway termini today. They would line the main commuter routes into the city, such as the Strand on Booth's map (see road running south-west to north-east at the bottom of Figure 6.6). Indeed, remnants of the role of the pub in historical commuting routes can still be seen on the periphery of cities, such as the many pubs situated along the Great North Road leading into London through Chipping Barnet. Many of the West London pubs, just as in the East End, were catering for people working locally (see the cluster around Covent Garden market immediately north of the Strand in the Figure 6.6). Other places had areas of exclusion determined by the land owner, so the district abutting the British Museum and Russell Square on Booth's map is markedly devoid of pubs, due to rules imposed by the area's land owner, the Duke of Bedford; an absence of pubs enhanced the value of the estate, although the map shows pubs situated on its margins, well positioned to supply the grand houses within the area. In fact, from the 1860s onwards a clear distinction could be made between pubs licensed for drinking on their premises and 'off-licenses', namely establishments where you could buy drink to consume elsewhere. There were many more of the latter in prosperous areas than in poverty areas.

Harrison's analysis shows that the dominance of pubs in poverty areas is not so much due to alcohol being a problem of poverty, but due to the importance of pubs for recreation for working-class people. Many pubs in these areas had licenses only for beer (that is, not for spirits); they provided a place of conviviality outside of the crowded home, as we saw in the analysis of the drink map of York in Chapter 4. The lit-up entrances to drinking establishments, as well as their ornate exteriors, were an attractive beacon in the dark streets of London, especially in the case of gin palaces (see Dickens quote above).

In addition to recreation, the pub served a variety of other purposes: as a meeting-place for public organisations of all types, whether reform movements or the emerging trade unions. Anne Kershen describes how in

1889 the strike committee of the Jewish branch of one of the trade unions, the AST (the Amalgamated Society of Tailors), met at 'the unimposing White Hart public house in Greenfield Street'.[48] Pubs were also used as the informal meeting point for jobbing tailors to pick up information on the availability of piecework; by the middle of the nineteenth century, rather than simply posting their names on lists of those seeking labour, the unemployed were themselves forced to wait in pubs in the hope of being selected for work by prospective employers.[49] Given that pubs provided an accessible location, with an interior hidden from the public eye, it is not surprising that they also served the criminal classes, with activities such as procuring prostitutes or gambling finding their natural place there.

Revisiting the juxtaposition of churches and pubs on the map of *Religious Worship . . . and Houses Licensed for the Sale of Intoxicating Drinks* it becomes clearer still the extent of the battle the temperance societies had to fight against the competing attractions of drinking establishments. Putting aside Anglican churches, which were more likely to have a dominant position on the street,[50] most religious institutions were in relatively tucked-away locations. More importantly, the social role of the pub, especially for the working man (and, to a lesser extent, the working woman) was as central to city life as its spatial position on the street. Nevertheless, as Harrison maintains, the pub and the temperance society were similar in their ability to provide a collective experience to the single working man, especially the many migrants from rural areas, integrating the newcomer 'into urban society by initiating him into business habits' and helping him to overcome the 'shock of transition' from the countryside to the city.[51]

The temperance societies in the United States had a similarly challenging role. In Chicago and other major cities the various temperance societies did battle with the large number of saloons on the city streets. Robert Graham's 1883 pamphlet, 'Liquordom in New York City, New York', contained a series of maps of sections of the city's streets, marked up with the various types of 'liquor' available. New York's infamous Bowery district is shown with an almost uniform array of saloons on every street in the area, although closer inspection reveals lager beer is present on many more streets than 'liquor' (spirits). The pamphlet's author Robert Graham was secretary of the Church of England Temperance Society, and had travelled to New York from England (via Canada) to organise temperance associations in the city. Having arrived in 1881 he was escorted on a tour of the city's 'slums' to gather statistics on liquor licenses in order to assess the 'state of temperance affairs in America'. His pamphlet compares the numbers of saloons, or drink-shops, and the churches and schools in New York City, writing that:

> It is an undoubted fact that just where the poverty and misery is greatest, there is the largest number of saloons. Granted squalid and overcrowded homes, with a minimum of comfort and a maximum of filth, it is not to be wondered at that saloons with polished woods, meretricious gilding, light, warmth, and freedom, should compete with and beat out of the field the three bare and comfortless rooms which are home only in name. To the real home in the city of New York, which is within the reach of every man in it, there can be no deadlier enemy than the 10,168 saloons which crowd its alleys and throng its courts.[52]

His pamphlet states that he drew the maps to emphasise 'the huge disproportion of saloons apparent to the eye . . . especially in the poorer quarters of the city'.[53] His conclusions were not dissimilar to those regarding London, highlighting the attractiveness of public drinking places for the deeply impoverished people of New York.

The second edition of his pamphlet, 'New York City and Its Masters', published in 1887,[54] highlighted the number of churches and schools (568) in comparison with the number of saloons (10,168), to show the 'startling disproportion between agencies for good and evil'.[55] The pamphlet's maps of the 23 assembly districts of the city contain statistics on the district's population as well as the number of churches, schools and saloons in the district, working out the ratio of saloons per head of population to argue for the need for a dramatic reduction in drinking establishments, especially in poverty areas. His work did not stop there. Graham went on to publish a pamphlet on 'Social Statistics of a City Parish', which investigated the social conditions, nationalities and so on of the population of Trinity Parish, New York, which was suggested as a template for further investigations into the situation in the city.

This was a period of intense activity in the social investigation of the alcohol problem.[56] Almost simultaneously with Graham's work, Henry Blair, a Republican senator from New Hampshire, was drawing up a map locating saloons across New York City to accompany his book *The Temperance Movement or the Conflict between Man & Alcohol*, published in 1888. His June 1886 enumeration of 10,168 saloons and places selling

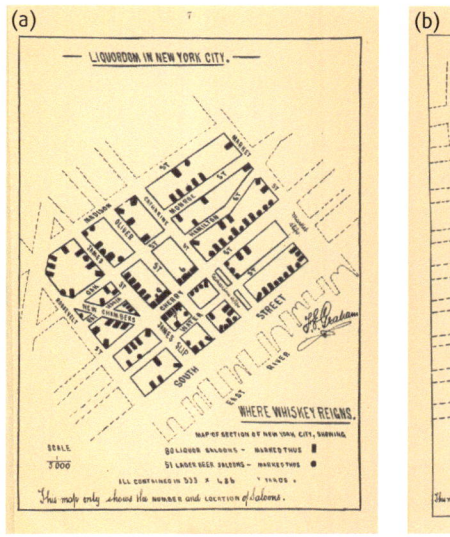
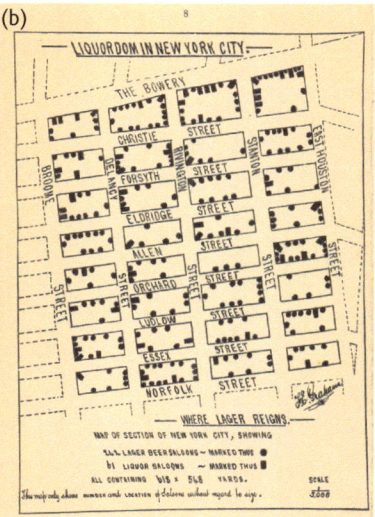
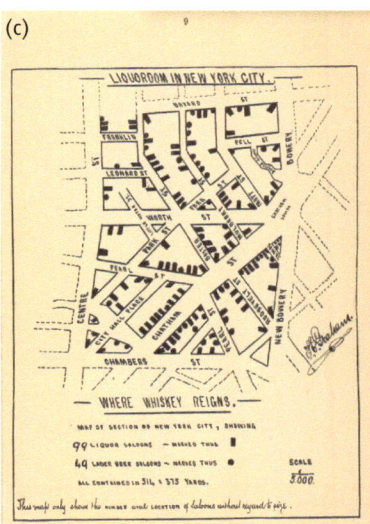
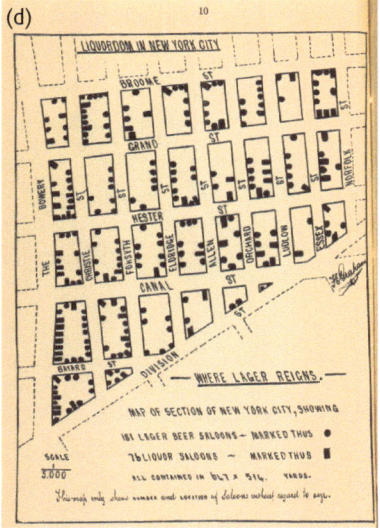

Figure 6.8 *Liquordom in New York City*, 1883.
Robert Graham, 1883. Copyright Cornell University – PJ Mode Collection of Persuasive Cartography.

alcohol within the metropolis, were charted on a map of 9,000 of these[57] that he termed as akin to a chart of despair:

> The eye is the chief inlet to knowledge, and the map of New York city which accompanies this book, upon which are located over 9000 of the 10,168 saloons and places where intoxicating liquor

was for sale in that metropolis on the thirtieth day of June 1886, looks like a chart of the capital city of the regions of despair. And when we consider that this great city controls the pivotal State of the Union, and how helplessly it drifts in the maelstrom of alcohol, we require more than the faith which removes mountains if we are still to hope for the republic.[58]

What is striking about Blair's map, 'of New York City from the Battery to Central Park, showing the Location of all Drinking-Places', is how it shows the combination of the spatial and social dynamics which were shaping the location of saloons in New York at the time. On the one hand, the map shows a dramatic decline in the number of saloons in the more prosperous, 'up-town' districts of Manhattan (the right-hand end of the map); the spatial distribution of drinking places seems to confirm the prosperity–poverty trajectory proposed by all the temperance writers. Yet closer examination shows a much more intricate pattern, with some streets carrying large numbers of saloons and others, many fewer.

One of the most interesting clusters is in a dense triangle of saloons sited on the edge of the Lower East Side, at the bottom of Bowery, a street lying in the shadow of the Third Avenue elevated railway, the 'El' (note the markings like railway tracks in Figure 6.10). Indeed, by the late nineteenth century the Bowery had become synonymous with squalor, with Theodore Roosevelt referring to it in an essay as 'a highway of seething life, of varied interest, of fun, of work, of sordid and terrible tragedy'.[59] Its reputation as a centre of prostitution and its all-night saloons were a matter of public concern, with commentators convinced that they were the source of the 'ruination of numberless men and boys'.[60]

In fact, if we zoom in on the Bowery, and its parallels such as Christie and Forsyth, it did indeed have many more saloons than elsewhere in the area. Further study shows that while there was a particularly large number of saloons around the convergence point of the three main routes in the district, the numbers reduced progressively with distance from the main routes, especially within the depths of the Lower East Side. There might be a social explanation for this: the Lower East Side was at the time the centre for Jewish immigrant settlement, and members of this community were much less likely to frequent bars than other poverty groups (see Figure 6.11). However, looking at Garyfalia Palaiologou's space syntax analysis of Manhattan c.1891, overlaid with the enlarged section of the saloon map in Figure 6.10, we can see that while there is a social logic to

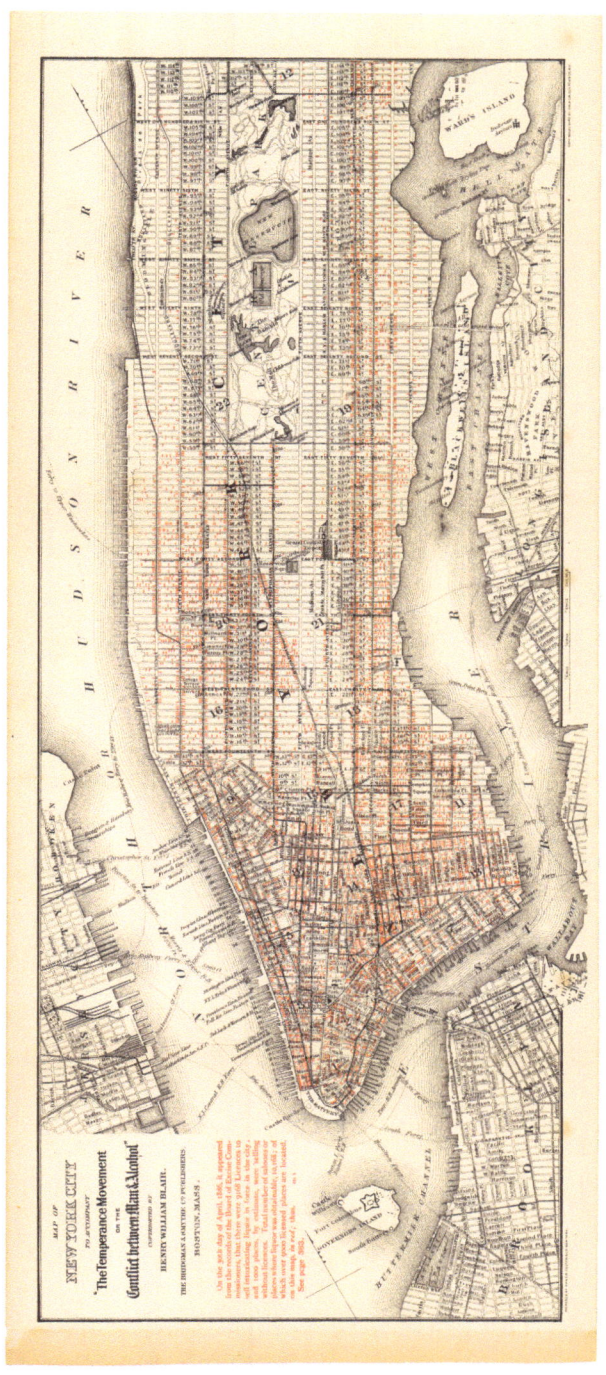

Figure 6.9 Saloon Map of New York City, 1888.

Henry Blair, 1888. Copyright Cornell University – PJ Mode Collection of Persuasive Cartography.

Figure 6.10 Detail of *The Temperance Movement or the Conflict between Man & Alcohol* (Saloon Map of New York City), 1888, overlaid on space syntax analysis of Manhattan c. 1891.

The section of the saloon map is highlighted with a black, dashed box.
Original map by Henry Blair, 1888, copyright Cornell University – PJ Mode Collection of Persuasive Cartography. Spatial syntax analysis by Garyfalia Palaiologou, 2017.

Figure 6.11 Detail from *A scene in the ghetto, Hester Street*, New York, c. 1902.

Photograph by Benjamin J. Falk, from Library of Congress Prints and Photographs Division Washington, D.C.

the distribution, it also follows a spatial logic: Christie and Forsyth streets (running south to north on the map) are both highly accessible streets at the neighbourhood scale; in other words, you would expect there to be much more passing traffic along those streets than on the streets lying to their east. On the other hand, when considering patterns of accessibility at a more local scale, the streets at the heart of Manhattan's Jewish quarter formed a localised area of relative inaccessibility, which would have helped create a sense of an inward-looking district that simultaneously connected outwards along its main roads.[61]

Temperance movements were much less active after the turn of the twentieth century. In the United States the passage of the 18th amendment to the constitution in 1919 brought into law the national prohibition against the manufacture of alcohol, which reinforced the anti-saloon legislation that had previously been passed in 1916 in nearly half of the states. Prohibition brought about a significant reduction in alcohol consumption and, in parallel, a rise in illegal production and consumption.

The need for temperance campaign maps reduced accordingly. The same was the case in other countries such as the United Kingdom, where the devastation of the First World War shifted the public focus onto legislation to alleviate ongoing problems with housing conditions, especially for soldiers returning from war. Yet social problems continued to manifest in cities and the worst of these, criminal activity, intensified in some areas – or at least drew greater public attention due to their concentration in certain inner-city areas.

Crime and deviance

Crime and deviance have been the backdrop of many of the maps we have seen so far, especially those concerning poverty. Maps of nationalities, race and religion have also mentioned the spatial incidence of crime; in some cases, such as the map of San Francisco's Chinatown, the authors have not been shy about attributing a rise in crime directly to racial factors.

It is no coincidence to find that both the maps presented below emanated from scholars affiliated with the University of Chicago. The notion of the city as a laboratory for studying urban society was first conceived by the group of sociologists based at the university who were known as the Chicago School. Robert Park, together with Ernest Burgess and colleagues, was one of the first to propose that the complexity of urban societies requires an empirical approach that controls the shape and form of the spatial environment as one might control a chemical in a laboratory.[62] The combination of their institutional backing and the setting of the rapidly growing city made Chicago the ideal place to 'do urban research' for much of the period leading up to the Second World War and on into the 1950s.[63]

One of the School's most influential ideas was Burgess's conception of cities as if they were made up of concentric zones. In an idea first articulated in 1925 in *The Growth of the City: An Introduction to a Research Project*, Burgess proposed that the growth of cities typically followed a concentric process of expansion: from an inner Zone I (termed Loop – clearly Chicago was the model) surrounded by a Factory Zone, set within Zone II (the Zone in Transition), surrounded by Zone III (the Zone of Workingmen's Homes), surrounded by Zone IV (a Residential Zone) and finally Zone V (the Commuters' Zone). Despite Burgess's explanation that his chart was 'an ideal construction',[64] its impact on planning ideas continues to this day. But another idea that appears in the same paper, disorganisation theory, is equally important for its influence on the maps that we will consider below (as well as in its long-term influence on the discipline of criminology).

Burgess argued that as cities expand, 'a process of distribution takes place which sifts and sorts and relocates individuals and groups by residence and occupation'[65] (see Figure 6.12). The result, he wrote, is a mosaic of social worlds comprised of immigrant areas such as Chinatown or the Jewish 'ghetto', whose inhabitants move progressively through each zone, seeking 'the Promised Land' beyond. These naturally evolved areas have developed alongside other residual areas, 'submerged regions of poverty, degradation and disease', where accepted rules of social behaviour are absent. If cities grow too fast, Burgess wrote, the internal movements of people through the zones create a 'tidal wave of innundation', leading to excessive social disorganisation in the form of crime, disorder, vice, insanity and suicide. He was stating that social disorganisation occurs where there is a lack of collective social values and

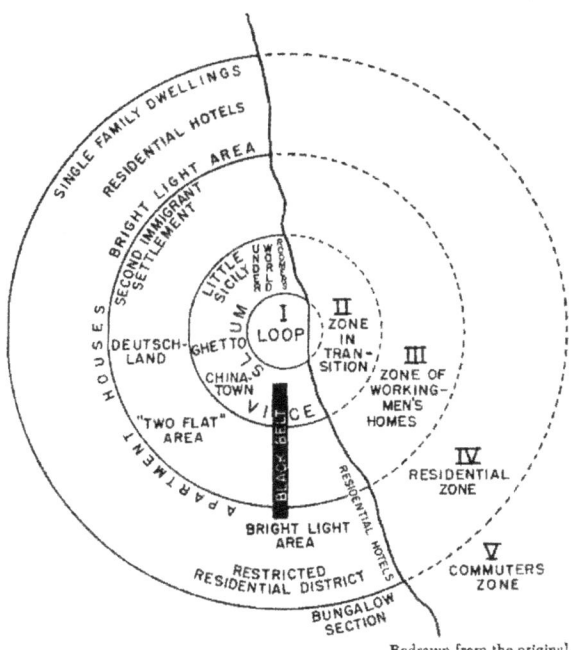

Redrawn from the original.

Figure 6.12 *Chart 2 – Urban Areas*, illustrating the growth of cities.

From E.W. Burgess, 'The Growth of the City: An Introduction to a Research Project,' in *The City*, ed. R.E. Park, E.W. Burgess and R.D. McKenzie (Chicago: University of Chicago Press, 1925), p. 38.

effective social control on deviant behaviour; so delinquency is the outcome of community breakdown, rather than individual deviance. Shaw's later work with McKay, which expanded the study to 21 American cities, supported Burgess's hypothesis that the physical deterioration of residential areas accompanied by social disorganisation is greatest in a central zone in the business district, and declines progressively from the inner city to its peripheral areas.[66] These propositions have since been refuted by scholars who argue that the model does not explain the reality of complex urban processes, yet the beauty of its simplicity means that notions of disorganisation – as well as the concentric zones model – continue to hold in many criminology studies today.

The following two maps show us different aspects of disorganisation at almost the same time, in 1920s Chicago. The first, *Map no. VII showing places of residence of 7,541 alleged male offenders placed in the Cook County jail during the year 1920, 17–75 years of age*, was created by a team headed by Clifford R. Shaw for the Local Community Research Committee of the University of Chicago. The map was published in the Committee's second monograph, *Delinquency areas: A Study of the Geographic Distribution of School Truants, Juvenile Delinquents, and Adult Offenders in Chicago*. Shaw and his colleagues' five-year study had considered the geographic distribution and the rate of occurrence of delinquency and crime, as a 'unique approach to the problem of conduct as it is influenced by the type of community background'.[67]

The map covers a vast territory (see Figure 6.13) as reflected in the great volume of statistics gathered for the study, which considered a variety of data sources on delinquency rates according to sex and age across various temporal periods. This was a sophisticated analysis. Instead of simply counting numbers by area, it took account of population size to calculate a rate by area.[68] The key findings were that delinquency rates varied in inverse proportion to distance from the city centre (namely, the Loop); that differences in delinquency rates differed according to the background of the community; and that especially high rates of delinquency were to be found in areas that were deteriorating physically. Notably, another aspect of the study's refinement was that it took account of change over time.

The study found that the addresses of male offenders were concentrated in the inner-city area (closest to the central-eastern section of the map). In fact, the area mapped by Hull-House a couple of decades earlier was quite close to the most densely marked area on the map (see map detail in Figure 6.14).[69]

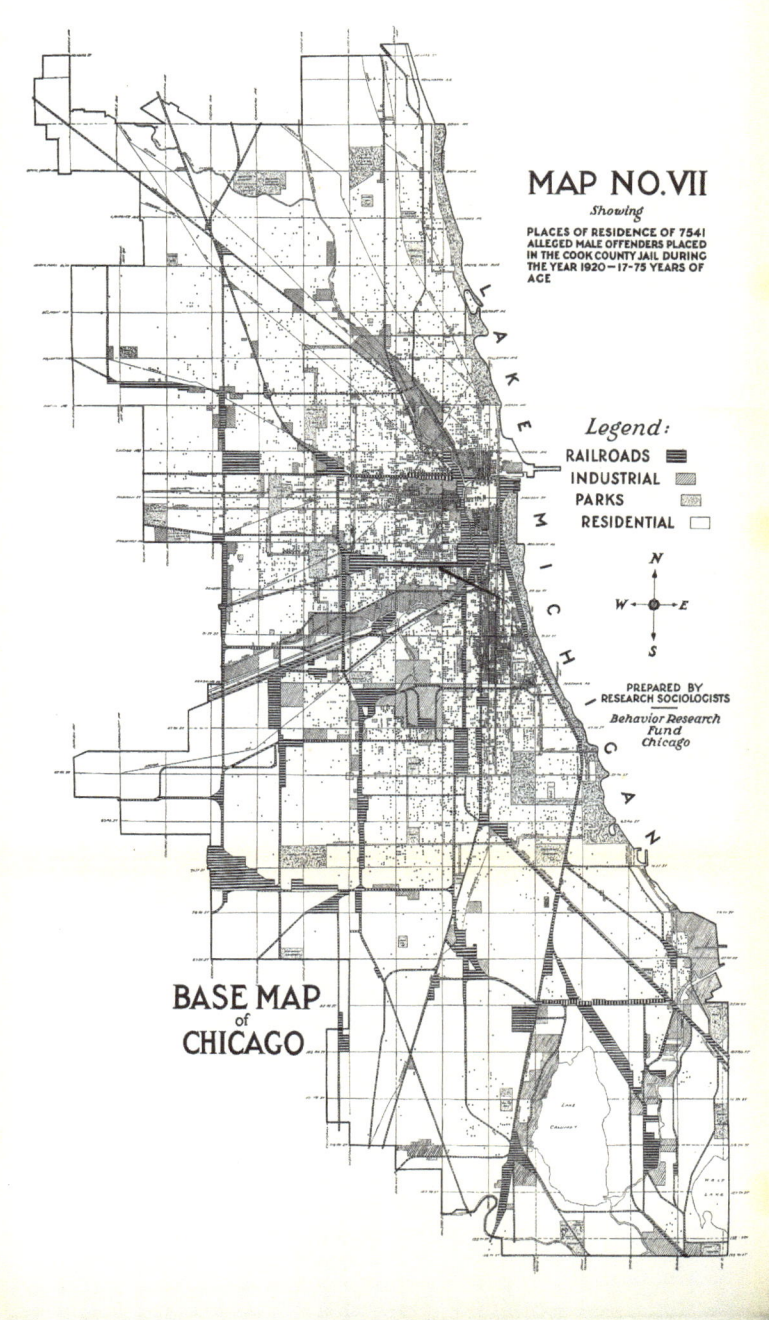

Figure 6.13 *Map no. VII* showing places of residence of 7,541 alleged male offenders placed in the Cook County jail during the year 1920, 17–75 years of age.

C.R. Shaw and F.M. Zorbaugh, *Delinquency areas: A Study of the Geographic Distribution of School Truants, Juvenile Delinquents, and Adult Offenders in Chicago* (Chicago: University of Chicago Press, 1929). Copyright University of Chicago Map Collection.

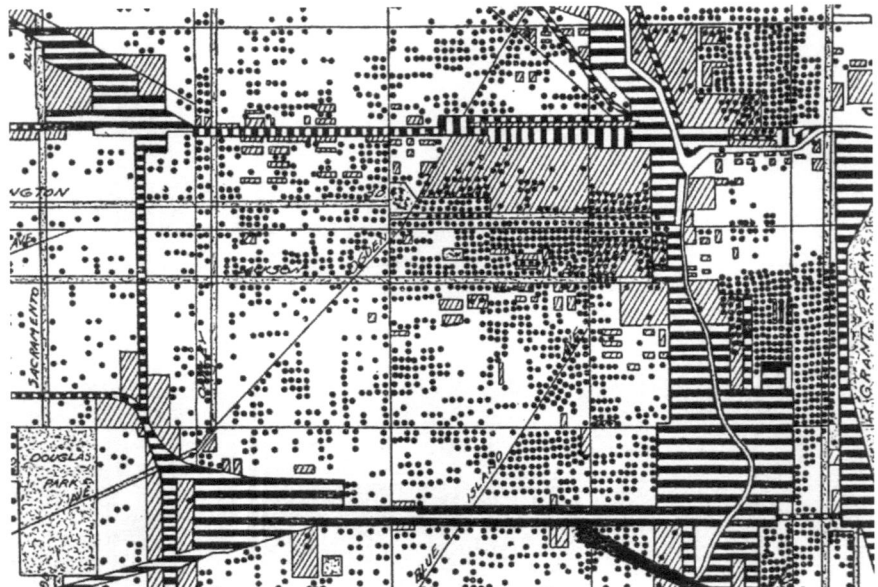

Figure 6.14 Detail of *Map no. VII*.
Shaw and Zorbaugh, *Delinquency Areas*. Copyright University of Chicago Map Collection.

Although Shaw and colleagues read the results as if they supported the concentric rings notion (with a marked decline in numbers as distance from the Loop increased), in fact, other patterns emerge from closer study of the map. For example, crime is greater in locations east of the river, especially in those further segmented by the railways. The text of the book covered many other topics (the map featured here was one of 10 large maps published in the book) and its conclusions also suggested that delinquency could be associated with the type of community within which people lived, stating that juvenile delinquents were influenced by their community's 'situation' – that is, the social environment within which young people grew up; again, this confirmed Burgess's propositions regarding social disorganisation.

A better-known text from almost the same time is *The Gang: A Study of 1,313 Gangs in Chicago* by Frederic M. Thrasher, published in 1927. Still in print today, the text recorded the youth gangs that had grown up in the city. His map outlined the typical areas of gangland, aiming to record their position in the life and organisation of the city. In addition, each dot on the map recorded the location of their meeting places, their 'favourite haunts and hang-outs'[70] (see Figure 6.15).

Figure 6.15 *Chicago's Gangland*, 1927.
From F.M. Thrasher, *The Gang*, 1927. Copyright Cornell University – PJ Mode Collection of Persuasive Cartography.

The text is full of the most intricate detail on the spatial nature of gang activity and the way in which disorderly behaviour takes place in the interstitial, marginal areas of the city, namely 'the spaces that intervene between one thing and another'. In fact, an interesting game could be played spotting the many synonyms for segregated areas used in the course of the book. The *wilderness*, the *slum*, the *colony* and the *territory* are found to nestle on the *barriers*, *borders* or *frontiers* of another gang's area. These are typically in low elevation areas – *valleys*, *gullies* or *canals* – which are segmented by railroad tracks or highways; in some cases, they occur in a veritable wilderness or so-called *blackspot*. Taking just one randomly selected page, we have a description of such an area:

> Across these turbid sewage-laden waters lie the crowded river wards. In the drab hideousness of the slum, despite a continuous exodus to more desirable districts, people are swarming more than 50,000 to the square mile. Life is enmeshed in a network of tracks, docks, factories and breweries, warehouses, and lumber-yards . . . ramshackle buildings . . . besmirched with the smoke of industry. In this sort of habitat, the gang seems to flourish best.[71]

As Thrasher himself explained, the interstitial area is akin to the natural phenomenon of 'foreign matter [which] tends to collect and cake in every crack, crevice, and cranny'. Similarly, he wrote, there are 'fissures and breaks in the structure of social organisation'[72] – which is how gang activity manages to survive: in the similarly fragmented 'interstitial region in the layout of the city'. Importantly, he refers both to local severances in the urban street network and to entire regions of disorganisation, such as the 'poverty belt', where neighbourhoods have deteriorated and populations are in flux or have abandoned the area entirely.[73]

The detail of the map in Figure 6.16, covers the same area as the detail of *Map no. VII* in Figure 6.14. It is interesting to note that Thrasher's 'Rooming House' area within the Loop was predicted in Burgess's paper, where he wrote that 'within a deteriorating area are rooming-house districts, the purgatory of "lost souls"'. Like many of the encircled areas on the map, this was not exclusively a gang area. It contained myriad other activities: churches, schools, clubs and banks and other 'wholesome institutions':[74] yet the Loop is said by Thrasher 'to form a sort of interstitial barrier between [it] and the better residential areas'.[75]

Spatial mechanisms of crime, deviance and immorality

Once the influence of the Chicago School started to wane, social scientists started to criticise the social ecological approach, stating that is was a form of environmental determinism. Many scholars continue to argue today that attributing any agency for architectural or urban settings to

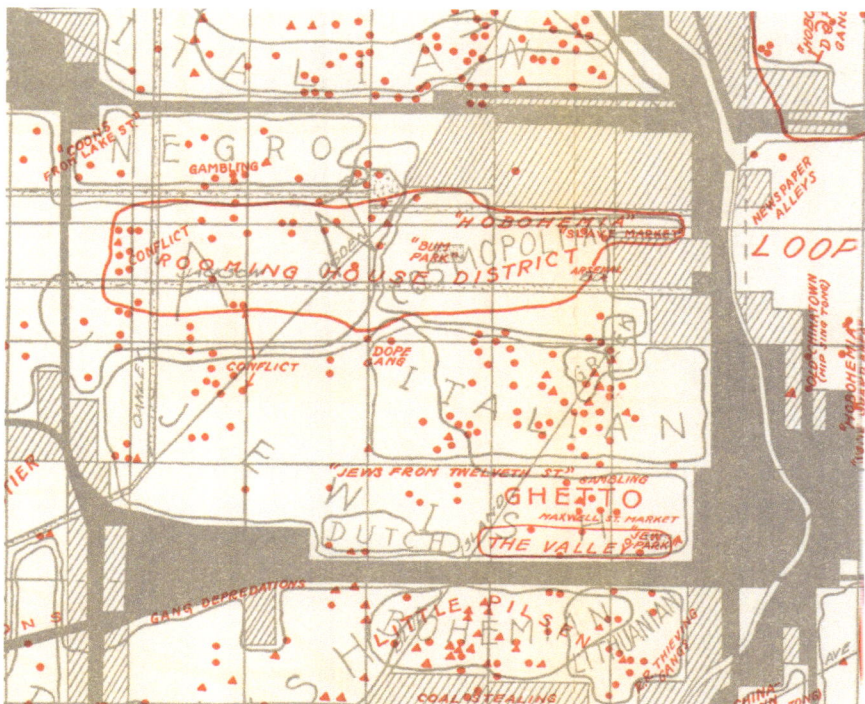

Figure 6.16 Detail of *Chicago's Gangland*, showing same area as Figure 6.14.
Thrasher, *The Gang*. Copyright Cornell University – PJ Mode Collection of Persuasive Cartography.

shape social outcomes 'may facilitate architectural thought but it is not a sociological analysis'.[76] Such concerns have emerged recently in several studies of the riots that took place in major cities in the United Kingdom in 2011. The research has been a subject of some controversy due to newspaper reports that a disproportionate amount of the arrested rioters had come from supposedly disorderly social housing estates.[77] In fact, a detailed spatial criminological analysis has found the only spatial association to be that rioters were more likely to engage in the disorder close to their home location and to select places visited frequently or at which they spent much of their time.[78] Rather than proving that bad design leads to crime occurring in specific types of locations, the studies found that certain types of crimes are shaped by spatial conditions, and that the built environment can create the preconditions for social deterioration and for crime to be more likely to occur in such areas.

The study of crime itself has shifted in recent times from a focus on criminal behaviour – on the psychological or social motivations for people to deviate from society's norms – to creating a sub-discipline of environmental criminology, which involves studying the circumstances in which crime takes place, such as the physical environment, the opportunities to commit crime, and the lack of safeguards to prevent the crime from taking place.[79]

Around the same time that studies of environmental criminology emerged, a new vision of how cities function was formulated in the field of architecture. Several propositions, such as the work of Kevin Lynch, and especially in the work of Hillier and Hanson, have argued for an approach that takes account of the relationship between the spatial configuration of cities and the way urban space is used in daily life.[80] Instead of modelling the city as a featureless plain in which crime takes place, scholars have started to examine the underlying processes that shape movement patterns in the city, the distribution of commercial and residential land uses across the street network, the relationship between street permeability and mixed uses, and the navigational and cognitive aspects of space, including topological and morphological properties of urban spaces. These studies have examined how urban spaces facilitate or impede social and economic processes. When looking at crime, they will typically consider the way the street network is organised and how both the movement densities and potential crime targets, in the form of land uses and people, are distributed across the network in ways that can have an effect on the distribution of crime patterns.

Nevertheless, there is currently insufficient integration between the built environment disciplines and crime science. While crime scientists may emphasise the influence of the built environment on crime, and increasingly employ sophisticated spatial statistical models in doing so, their analysis of urban form and layout tends to be inconsequential. It is not unusual to find, for example, that crime data are summarised by area, or measure distances 'as the crow flies', even though the spatial constitution of the street layout can have an important role in where crime takes place – and where it does not. For built environment-based research, the reverse has frequently been the case. As a result, criminologists can be sceptical about the results from studies of the relationship between street networks and crime, although there is a growing cohort of scientists whose use of spatial mapping coupled to robust statistical analysis of crime is attracting attention in the field.[81]

One key aspect of the latest crime mapping research is the increasing ability to go beyond identifying clusters or hot spots of criminal activity towards analysing the common factors between different locations of crime. For example, are burglaries more likely to take place in dead-end streets than in streets with passing traffic? Does it matter if the dead-end is visible from the main road or not, and so on? We are currently at a juncture where increasing computer power, alongside a growth in websites where people can check on their own local patterns of crime, give the impression that the ability to study crime is accessible to any person with a decent computer – a form of data democracy that is unprecedented. As we will see in the next chapter, this may very well be a false vision. In many ways, 200 years of social cartography have brought us much closer to understanding the spatial logic of urban societies, but in many other ways the ability to map data speedily and in a highly detailed fashion increases the responsibility of the spatial scientist to explain social structures more comprehensively than ever before.

Notes

1. Booth, 'Poverty Series Survey Notebooks (Online Archive)', BOOTH/B/B351, pp. 126–7.
2. R.J. Sampson and S. Raudenbush, 'Seeing Disorder: Neighborhood Stigma and the Social Construction of 'Broken Windows', *Social Psychology Quarterly* 67, no. 4 (2004).
3. The quote is from one of Booth's notebooks, describing Dorset Street – 'the worst street I have seen so far' – which became notorious as the site of the murder of Mary Kelly, the last of Jack the Ripper's victims. Booth, 'Poverty Series Survey Notebooks (Online Archive)', BOOTH/B/B351, pp. 100–1.
4. T. Osborne and N. Rose, 'Spatial Phenomenotechnics: Making Space with Charles Booth and Patrick Geddes,' *Environment and Planning D: Society and Space* 22 (2004), p. 216.
5. F. Driver, 'Moral Geographies: Social Science and the Urban Environment in Mid-Nineteenth Century England,' *Transactions of the Institute of British Geographers* 13, no. 3 (1988).
6. My italics. Quotation is from Joseph Fletcher, Secretary to the London Statistical Society in 1849, quoted in Driver, 'Moral Geographies,' 279.
7. Evans, 'Rookeries and Model Dwellings', p. 95.
8. Evans, 'Rookeries and Model Dwellings', p. 95, citing W. Beckett Denison, 'On Model Lodging Houses', 1852.
9. Evans, 'Rookeries and Model Dwellings', p. 96.
10. Driver, 'Moral Geographies,' pp. 282–4.
11. J.J. Varga, *Hell's Kitchen and the Battle for Urban Space: Class Struggle and Progressive Reform in New York City, 1894-1914* (New York: Monthly Review, 2013).
12. Varga, *Hell's Kitchen*, p. 101.
13. R. Park and E. Burgess, *The City* (Chicago: University of Chicago Press, 1925), p. 21, cited in M. Valverde, *Law and Order: Images, Meanings, Myths* (Abingdon and New Brunswick, NJ: Rutgers University Press, 2006), p. 142.
14. See detailed analysis of this subject by B. Campkin, *Remaking London: Decline and Regeneration in Urban Culture*, vol. 19 (London and New York: IB Tauris, 2013).
15. The Hebrew bible's book of Joshua tells the story of Rahab the prostitute (who gave refuge to Joshua's spies), whose house was built into the wall of the city of Jericho. [*Joshua* 2:1–24].
16. Geremek and Birrell, *The Margins of Society*. See in particular the map of 'Prostitution in Paris 1380' on page 88.
17. Geremek and Birrell, *The Margins of Society*, p. 87.
18. Some of London's old City parishes were even named in this way, so that there were two wards: Farringdon *Without* and Farringdon *Within*, set either side of the city's walls. See also L. Vaughan, 'The Paradox of City Walls: Enclosure, Boundary, Barrier,' *Lobby magazine* 3 (2015).

19. G. Palsky, 'Connections and Exchanges in European Thematic Cartography. The Case of 19th Century Choropleth Maps,' *Belgeo. Revue belge de géographie*, no. 3–4 (2008).
20. Friendly and Denis, 'Milestones in the History of Thematic Cartography.'
21. Palsky, 'Connections and Exchanges,' p. 413.
22. A.J.B. Parent-Duchâtelet, *On Prostitution in the City of Paris. From the French of M. Parent Duchâtelet*, second ed. (London: T. Burgess, 1840).
23. P. Beirne, *Inventing Criminology: Essays on the Rise of 'Homo Criminalis'* (Albany: State University of New York Press, 1993), pp. 128–9.
24. M. Ryan, *Prostitution in London: With a Comparative View of That of Paris and New York, as Illustrative of the Capitals and Large Towns of All Countries: And Proving Moral Deprivation to Be the Most Fertile Source of Crime and of Personal and Social Misery: With an Account of the Nature and Treatment of the Various Diseases Caused by the Abuses of the Reproductive Function* (London: H. Bailliere, 1839), p. 73. Alongside its counterpoint, a map of the origins of prostitutes provided to the reader a statistical picture of migration from the countryside and its association with vice at that point in time, raising numerous questions regarding whether the women brought the immoral behaviour with them, or whether it emerged as a result of the conditions of their new environment.
25. Ryan, *Prostitution in London*, p. 74.
26. Ryan, *Prostitution in London*, p. 219. Here the author is describing prostitution in America.
27. Farwell, *The Chinese at Home and Abroad*, p. 39.
28. Booth, 'Poverty Series Survey Notebooks (Online Archive)'. BOOTH/B/B351 pp. 104-5.
29. Booth, 'Poverty Series Survey Notebooks (Online Archive)'. BOOTH/B/B351 pp. 131–3. 'Pink, on map purple'; 'lb to purple, d blue in map', means in other words that the street has improved from purple to pink or from dark blue to light blue or purple in the past decade; 'lb. db. in map' means that the street has risen from dark blue to light blue over time.
30. W.S. Robinson, *Muckraker: The Scandalous Life and Times of W.T. Stead: Britain's First Investigative Journalist* (London: Robson, 2012).
31. S. Wood-Lamont, 'W.T. Stead's Books for the Bairns,' http://www.attackingthedevil.co.uk/worksabout/bairns.php.
32. Robinson, *Muckraker*.
33. W.T. Stead, *If Christ Came to Chicago – a Plea for the Union of All Who Love in the Service of All Who Suffer* (London: The Review of Reviews, 1894), p. 117. Stead was also confident that maps could play a role in solving a problem. Elsewhere in the book he advocated that the city's Central Relief Committee should issue a map highlighting districts requiring 'visitation and relief', which ideally should correspond to electoral districts, all the better for relieving the distress of the city's citizens (see p. 139).
34. Ryan, *Prostitution in London*, p. 75.
35. M. Douglas, *Purity and Danger: An Analysis of Concepts of Pollution and Taboo* (London: Routledge and Kegan Paul 1966). Uncleanness is more akin to the way in which the ancient Israelites maintained boundaries through the rules of kashrut (continued today in orthodox Jewish practice). Order is created by classifying animals and the line of the social boundary is held by strict rules regarding separation between communities. In this way contamination or pollution are avoided. See commentary on Douglas's theories in Sibley, *Geographies of Exclusion*, pp. 36–7.
36. P. Hubbard, 'Red-Light Districts and Toleration Zones: Geographies of Female Street Prostitution in England and Wales,' *Area* 29, no. 2 (1997), p. 133–5.
37. Hume, *Condition of Liverpool*, p. 22.
38. The American Society for the Promotion of Temperance, for example, was founded in 1826. The British Association for the Promotion of Temperance was established by 1835, although treatises against alcohol were published many decades earlier. Both movements probably came about due to a growing recognition that excess consumption of alcohol was dangerous; in fact poisonous. See S. Couling, *History of the Temperance Movement in Great Britain and Ireland* (London: W. Tweedie, 1862).
39. J. Kneale, 'The Place of Drink: Temperance and the Public, 1856–1914,' *Social & Cultural Geography* 2, no. 1 (2001), p. 53.
40. J. Kneale and S. French, 'Mapping Alcohol: Health, Policy and the Geographies of Problem Drinking in Britain,' *Drugs: Education, Prevention and Policy* 15, no. 3 (2008), p. 234. Once drinking shifted back to being viewed as a medical problem, pub maps declined rapidly in number. See more on contemporary geographies of drinking in Kneale's article.

41. C. Dickens, *Sketches by Boz – Complete in One Volume* (Paris: Baudry's European Library, 1839), pp. 143–4.
42. Dickens, pp. 146. Lethe-water probably alludes to 'Lethe', one of the rivers of Hades, drinking of which is said to induce oblivion. The decline of the countryside surrounding London was similarly symbolised by the transformation of a country pub into 'a mere suburban gin-palace'. See E. Walford, *Hampstead: Rosslyn Hill, vol. 5: Old and New London* (London: Cassell, Petter & Galpin, 1878), British History Online, http://www.british-history.ac.uk/old-new-london/vol5/pp483-494.
43. See more information on this map on the Cornell University website at https://digital.library.cornell.edu/catalog/ss:19343434.
44. Booth, *Life and Labour of the People in London*, pp. 119–20.
45. This is from a section on Hackney, close to the River Lea; Booth, *Life and Labour of the People in London*, p. 338. For Booth's police officers, the connection between intoxication and crime was obvious: 'As to warning publicans not to serve men on the verge of drunkenness, Drew said magistrates now thought it was the police's duty as much to prevent as to detect crime, therefore they had to warn Publicans. But he admitted it was a counsel of perfection & in practice was not often carried out.' Booth, 'Poverty Series Survey Notebooks (Online Archive)', BOOTH/B/350, p. 113.
46. The map was entitled *Map Showing Number of Public Houses in a District of Central London*. Joseph Rowntree was, aside from being the founder of a highly successful chocolate business, father to Seebohm Rowntree, author of the study of poverty in York seen in Chapter 3. His interest in temperance emerged in the late 1880s following a talk at the Yorkshire Quarterly Meeting of Quakers.
47. Harrison, 'Pubs,' p. 161.
48. Kershen, *Uniting the Tailors*, p. 136.
49. Kershen, *Uniting the Tailors*, p. 5.
50. See aforementioned analysis, comparing churches and synagogues, in Vaughan and Sailer, 'Metropolitan Rhythm of Street Life'.
51. Harrison, 'Pubs,' p. 184–5.
52. Graham, cited in P.T. Winskill, *The Temperance Movement and Its Workers: A Record of Social, Moral, Religious, and Political Progress* (London, Glasgow, Edinburgh and New York: Blackie, 1892), p. 51.
53. R. Graham, 'New York City and Its Masters,' ed. Church Temperance Society of New York (New York: Church Temperance Society's Offices, 1887), p. 6.
54. Graham, 'New York City and Its Masters.'
55. Graham, 'New York City and Its Masters,' p. 11.
56. Other maps include *The Lighthouse Saloon Map of Philadelphia*; this was published in 1901 in a pamphlet that described the operations of the 'Lighthouse' mission during the preceding five years.
57. The map states that 'On the 30th day of April 1886, it appeared from the records of the Board of Excise Commissioners, that there were 9168 Licenses to sell intoxicating liquor in force in the city, and 1000 places, by estimate, were selling without license. Total number of saloons or places where liquor was obtainable, 10168; of which over 9000 licensed places are located on this map.'
58. H.W. Blair, *The Temperance Movement or the Conflict between Man & Alcohol* (Boston: William E. Smyth Co, 1888), p. 363.
59. T. Roosevelt, *History as Literature* (New York: Charles Scribner's sons, 1913), VII: Dante and the Bowery, http://www.bartleby.com/56/7.html.
60. O.L.P, 'Letter to the Editor: An All-Night Saloon,' *New York Times*, October 23, 1910.
61. The analysis of the street layout of Manhattan at the scale of the entire island finds the main north–south routes being the most accessible; at the more local, neighbourhood scale (segment angular integration at radius 2500, illustrated in the main space syntax image in Figure 6.10), it is avenues such as the Bowery, running on to become Second Avenue, that are the most likely to have generated high rates of pedestrian movement alongside significant vehicular transport. The inset of 6.10 shows analysis of the same measure at the radius of 800m, a scale which predicts local movement flows. Full space syntax analysis of Manhattan over time can be found in G. Palaiologou, 'Between Buildings and Streets: A Study of the Micromorphology of the London Terrace and the Manhattan Row House 1880–2013' (Ph.D. diss., UCL, 2015).

62. Park and Burgess, *The City*; R.E. Park, 'The Urban Community as a Spatial Pattern and a Moral Order,' in *The Urban Community*, ed. E.W. Burgess (Chicago: University of Chicago Press, 1926); R.E. Park, 'The City: Suggestions for the Investigation of Human Behavior in the City Environment,' *American Journal of Sociology* 20, no. 5 (1915); R.E. Park and E.W. Burgess, *Introduction to the Science of Sociology* (Chicago: University of Chicago Press, 1921).
63. T.F. Gieryn, 'City as Truth-Spot: Laboratories and Field-Sites in Urban Studies,' *Social Studies of Science* 36, no. 1 (2006, February 1), p. 5.
64. E.W. Burgess, 'The Growth of the City: An Introduction to a Research Project,' in *The City* ed. RE Park, et al. (Chicago University of Chicago Press, 1925), p. 36.
65. Burgess, 'The Growth of the City,' p. 38.
66. C.R. Shaw and H.D. McKay, *Juvenile Delinquency and Urban Areas: A Study of Rates of Delinquency in Relation to Differential Characteristics of Local Communities in American Cities* (Chicago: University of Chicago Press, 1942).
67. University of Chicago, 'Back Matter,' *Social Service Review* 4, no. 4 (1930).
68. Nevertheless, although there were a large number of areas (113), these areas were large; the researchers combined census tracts to create areas of a square mile in size, and even larger areas on the city's outskirts. E.H. Sutherland, 'Delinquency Areas: A Study of the Geographic Distribution of School Truants, Juvenile Delinquents, and Adult Offenders in Chicago. Clifford R. Shaw, Frederick M. Zorbaugh, Henry D. Mckay, Leonard S. Cottrell (Review),' *American Journal of Sociology* 36, no. 1 (1930).
69. Hull-House was on Halstead, which runs north–south, two grid squares in from the left, just north of the area marked in black/white horizontal stripes, which by this time was covered by railway tracks.
70. Unsurprisingly, Thrasher was familiar with Shaw's work and references some of his texts on delinquency, including 'Delinquency Areas'.
71. F.M. Thrasher and J.F. Short, *The Gang: A Study of 1,313 Gangs in Chicago* (Chicago: University of Chicago Press, 1963 (first published 1927), p. 9.
72. Thrasher and Short, *The Gang*, p. 22.
73. Although not central to the discussion here, it is interesting that Thrasher describes gangs themselves as being interstitial in the lifespan of a boy's life, between childhood and maturity (or marriage), supplying activities ideal for the spare time (an interstitial temporality) of an adolescent boy. Thrasher and Short, *The Gang*, pp. 36–7.
74. Thrasher and Short, *The Gang*, p. 5.
75. Thrasher and Short, *The Gang*, p. 7.
76. H. Gans, 'Some Problems of and Futures for Urban Sociology: Toward a Sociology of Settlements,' *City & Community* 8, no. 3 (2009), p. 216.
77. R. Ball and J. Drury, 'Representing the Riots: The (Mis)Use of Figures to Sustain Ideological Explanation,' *Radical Statistics*, no. 106 (2012).
78. P. Baudains, A. Braithwaite and S.D. Johnson, 'Target Choice During Extreme Events: A Discrete Spatial Choice Model of the 2011 London Riots,' *Criminology* 51, no. 2 (2013). This finding is similar to the space syntax analysis of the riots, which found a large proportion of arrests took place in town centres and within a five-minute walk from social housing estates. B. Hillier, 'Credible Mechanisms or Spatial Determinism,' *Cities* 34 (2013).
79. P.J. Brantingham and P.L. Brantingham, *Environmental Criminology* (Beverly Hills and London: Sage Publications, 1981).
80. K. Lynch, *The Image of the City* (Cambridge, MA: MIT Press, 1960); Hillier and Hanson, *The Social Logic of Space*.
81. Recent work has attempted to bridge the two gaps, such as the work in crime science by Shane Johnson on burglary and the work in space syntax by Lusine Tarkhanyan on drug crime. L. Summers and S.D. Johnson, 'Does the Configuration of the Street Network Influence Where Outdoor Serious Violence Takes Place? Using Space Syntax to Test Crime Pattern Theory,' *Journal of Quantitative Criminology* 33, no. 2 (2017); S.D. Johnson and K.J. Bowers, 'Permeability and Burglary Risk: Are Cul-De-Sacs Safer?,' *Journal of Quantitative Criminology* 26, no. 1 (2010); L. Tarkhanyan, 'Drug Crime and the Urban Mosaic: The Locational Choices of Drug Crime in Relation to High Streets, Bars, Schools and Hospitals,' in: *9th International Space Syntax Symposium*, ed. Y.O. Kim (Chair), H.T. Park and K.W. Seo, 101:1–101:13 (Seoul, South Korea: Sejong University, 2013).

7
Conclusions

> The truth is that the dynamic of urbanism as we know it makes inevitable the syndrome of violence, alienation, high crime rates and delinquency that we associate with our cities. Once they have grown too big, these problems are unavoidable.[1]

The essential spatiality of urban social phenomena

Many of the maps shown in this book have valency in communicating statistics about society. They have used maps as means of capturing data gathered from direct observation – and latterly, surveys – in their spatial setting: using the mapped structure of urban form to classify the data spatially. In some cases, they have changed thinking about poverty, disease, segregation and crime, by highlighting the problems associated with these phenomena, and by locating their causes. We have also observed how space syntax can deepen the understanding of the relationship between spatial layout and social problems, by providing the tools to describe and analyse street networks empirically.

We have seen for example evidence to support Charles Booth's observations regarding the impact of the city on the long-term impoverishment of its inhabitants. While there is clearly not a one-to-one relationship between society and space, poverty is manifestly bound up with spatial isolation. More subtle socio-spatial findings, such as the role of urban streets in creating opportunities for encounters between people from different backgrounds, have also been shown to be fundamental to explaining how some groups have managed to overcome their apparent social segregation: just as deprivation is a multi-dimensional phenomenon that can worsen if it continues over time, so are other social conditions: one's membership of a minority religion need not preclude social integration in other settings, so long as the environment allows for different degrees of spatial integration to be maintained. Similarly, sub-cultures of fashion, politics or economic activity will exploit the urban setting, so long as it is not too sharply divided. Disease and crime are also frequently associated with spatial effects.

We need to keep Booth's example in mind when developing more informative definitions of deprivation for tackling urban problems comprehensively. To a certain extent this is understood. Internationally, the Millennium Development Goals propounded to 'make poverty history' initially comprised 8 goals and 21 targets with indicators, in a recognition that poverty is multi-dimensional.[2] In the United Kingdom the Index of Multiple Deprivation has a similar conception, defining deprivation as comprising many interconnected factors additional to income, such as health deprivation, environmental conditions and barriers to housing.

Cities sort people through property and labour markets, through social organisations, but also through their patterns of spatial organisation. The results of these dynamics of mixing are the spatial patterns of micro-segregation of people, land uses and activities. Cities are shaped simultaneously by built form, namely the sum total of buildings linked together by public space, as well as the complex system of human activity linked together by social interaction:

> The social city is either side of the physical city: it brings it into existence, and then acts within the constraints it imposes. It seems unlikely that either is a wholly contingent process. But both relations raise uncomfortable issues of determinism: how can a physical process in the material world relate to a social process in a non-trivial yet systematic way. This places philosophical as well as methodological obstacles in the path of reflection and research.[3]

In effect, the myriad maps presented in the preceding chapters are transcriptions of different geographies of inequality, made distinctive by the specific socio-spatial dynamics at play in their localities. By viewing the maps as historical records of these geographies, it has been possible to start to arrive at a clearer understanding of both the social and the political context in which they were created, but also at a more general understanding of how cities work as social systems.

In the case of disease, for example, it has been clear how a combination of poor housing, overcrowding, a lack of sanitation and general social neglect can, along with natural features of the area, exacerbate the numbers of people falling ill from certain diseases. The striation of cities by railway or highway infrastructures can compound social problems on the ground, making it more difficult to obtain work, to gain access to healthy food or more generally to integrate poor and rich in the same location. As mentioned already, disease mapping remains a vital tool in pinpointing loci of disease, but also in investigating underlying spatial or environmental causes of disease. Research might enquire as to whether the terrain, accessibility and access to healthy food collectively influence patterns of obesity and diseases related to this. The fact that poor health and poverty are still inexorably intertwined is evidently partly to do with poor housing, but there are many more factors that require investigation. In some cities diet-related diseases are increasingly being associated with access to fast food that is cheap but lacking in nutrition.

The chapters on poverty maps showed that deprivation is the outcome of myriad contributory factors, including social conditions, regularity of income and wider spatial accessibility, which are collectively brought to bear on long-term social outcomes. In addition, the way in which street networks are organised both locally and city-wide can bring about greater or lesser opportunities for wealth and poverty to be close to each other. What is clear is that interventions need to be made in the social causes as well as the spatial effects of poverty.

The historical maps of race and ethnicity have highlighted a different aspect of urban society. They have shown how frequently the spatial clustering of minority groups can be perceived as problematic because of the size of the settlement, or because of underlying issues of racism or at least ethnic marginalisation via the attitudes of the majority. Poverty coupled to ethnic clustering is frequently a contributory factor in the long-term stagnation of a group in an area, especially when the underlying cause is in fact spatial segregation – though equally a group may stay in an area due to benign factors, such as the presence of religious or cultural institutions, or commercial premises. The reality is that there are almost as many factors involved in minority ethnic clustering as there are maps recording it; so, the map of black settlement in Philadelphia would need to consider the group's political history as well as its poverty situation and state of origin, while the map of Jewish settlement in London shows the importance of accounting for how long a group has been in the country and for patterns of work as well as country of origin.

Once spatial isolation is involved, it is likely to be a contributory factor in whether the cluster is a ghetto or an ethnic enclave, whether it persists or is just a station-point on the way to acculturation or integration. Taking these considerations together, we can see that simply labelling a group as segregated disregards a highly complex socio-spatial constellation.

From the point of view of crime, the review in Chapter 6 found that while it is viewed as being disproportionately tied up with poverty many crimes in fact have fundamental environmental conditions, with certain anti-social behaviours or crimes gravitating to the more isolated areas of the city, while crimes that involve preying upon people (such as robbery or pickpocketing) have their own spatial demands. Nowadays there is much greater understanding about the socio-spatial mechanisms involved in where crime occurs, but there is plenty more to investigate in the burgeoning field of environmental criminology.

What is evident is that maps of crime, poverty or disease can reveal spatial regularities that underpin many of the most urgent social problems present in cities today. Cities continue to be characterised by segregation of different forms: enclosure and exclusion, and increasingly, social, economic, and political fragmentation.[4] The massive growth of the urban population across the globe is creating new forms of urban inequality on a sharper and larger scale than ever before. It is also evident how cleavages in terms of socioeconomic status, lifestyles and cultural and ethnic identities correspond to phenomena of residential segregation, so that in their more extreme manifestations the result is a form of hyper-segregation that combines both poverty and race. This trend toward new forms of spatial segregation has meant that some authors have forecasted that in the future cities will evolve into fragmented patchworks of impoverished ghettoes and affluent enclaves.[5]

Social mapping has different challenges nowadays than in the past. We could not draw up a Booth map today as we do not have access to the level of detail on the population, although we can get close to it. However, issues with the quality of data continue. These are due now not so much to an absence of data as to an abundance of it, giving the impression that we can be all-seeing and all-knowing about society today. Yet this is patently untrue. If we take patterns of employment, for example, it is almost impossible to grasp the full extent of a labour market, given that statistics rarely include self-employed and casually employed people, whose earnings change daily, let alone people working illegally. Similarly, it is very difficult to capture data on where work is being carried out, especially if it is in informal locations, such as people's front rooms, or above or behind high-street shops. National statistics rarely record this sort of information effectively, which means that it is overlooked in drawing up policy and then also undervalued when drawing up plans for urban areas.

Bearing this in mind, it is important to note those urban theorists who have asserted the essential spatiality of urban social phenomena. For example, when Ed Soja claimed in *Seeking Spatial Justice* that the social problems common to the Parisian *banlieues* were the result of situating new immigrants at the city's edge, he was highlighting the vital role urban space can play in shaping social patterns.[6] Others have noted that social phenomena need to be read as inherently spatial phenomena to be fully understood; the '"ghetto" is . . . [after all] an empirically determined, physical, quantifiable, experiential object'.[7]

Rethinking urban social problems spatially

Given the complex interrelationships between the configuration of local streets and wider cross-city (as well as global) trajectories it is no surprise to find that urban society evolves in a complex fashion. Yet this complexity realises itself in measurable spatial patterns, which differ due to local circumstances.

This book's opening chapter highlighted the historical importance of the Venice ghetto, not only because of its being the prime example of enforced isolation, but also because a close reading of its spatial nature reveals so much of the complexity of urban social problems. History shows that the position of the Venice Ghetto, a peripheral island that was nevertheless situated within a few turnings from the main thoroughfare, meant that it was possible for its inhabitants to have high rates of commercial and cultural interchange with Venetian society, despite their spatial confinement. Ironically, as Richard Sennett has noted, spatial isolation led to a flowering of a 'common culture' between the wide variety of Jewish sects and nationalities who, having arrived from elsewhere in the world, found themselves together for the first time in a single place.[8]

Physical and social dynamics of public space play a central role in the formation of publics and in the public culture, and although the Jewish inhabitants of Venice (and subsequently of Rome) trod a hazardous pathway through the city as they carried out their daily lives, especially when demands of religion required them to move in groups (such as for funerals), they were able to shape their opportunities for interaction with the world at large while traversing the city's streets and canals. In addition, notwithstanding the constraints on their movement at night, the ghetto also provided an opportunity to create a state of mutual solidarity, as well as a place of refuge for Jewish inhabitants at times of insecurity, such as at Easter 1766 in Rome:

> The Jews in this city are indulged in the use of synagogues; but are obliged to live all together in the Ghetto, as they call such places in the cities of Italy. At nine o'clock every evening the gates of the place where they live are shut up, and opened again in the morning; but at Easter they are locked up from Thursday in passion-week 'till the Monday following, during which time no Jew dares to be seen abroad. When they appear in the streets they are distinguished by a piece of yellow silk, or crepe, on the crown of their hats and are subject to a great penalty if seen without it. They are most of them very poor, and little respect is paid to the riches of them. Their synagogue has a mean appearance; yet it has some fine apartments, adorned with a great number of silver lamps.[9]

It is difficult of course to know for sure if the opportunities for encounter that arise from the historical arrangement of streets were always realised. The literature on daily encounters in the public realm tends to be divided on whether physical co-presence in the public realm necessarily translates into meaningful face-to-face interaction or if it remains superficial and at the level of familiarity. The serendipity of casual encounter that arises from the density of people present in cities can be linked to the historical development of an urban civic culture, yet in the early days of the modern city, theorists such as Simmel were concerned that the move to the city made relationships remote and encounters anonymous.[10] Possibly a more useful approach is to see these matters as being on a continuum, from casual familiarity due to physical co-presence, to encounter, to interaction and – occasionally – to actual social engagement. As such, public space without public presence is unarguably dysfunctional and, in contrast, public space becomes most meaningful when it encourages an encounter between people from different backgrounds. Objectively speaking, situations where people were physically isolated in cities was (and is) likely to have created the conditions for their social isolation.

What is evident from many of the historical cases that we have seen is how a normative mixing of people from different backgrounds within the public realm was not necessarily seen as problematic, but higher concentrations of visibly distinctive people from a different cultural or religious background were much more likely to bring about a degree of fear or hostility. Although many of the disease maps we saw in Chapter 2 would effectively mirror maps of poverty, it has been clear how the juxtaposition of racial segregation, poverty and disease (or at least fear of disease) led to the physical isolation of the Chinese inhabitants of San Francisco and its many sister settlements. Physical isolation helped intensify racialised stereotyping, such that Farwell's text that accompanied the 1885 maps had sentences that would not bear repeating nowadays. In other words, when physical isolation is combined with a lack of intergroup contact, divisions between hostile communities can only intensify. Even nowadays, cities such as Belfast, where the divisions along 'peace lines' have created sharp boundaries between national-religious groupings, feature places that become perceived as 'out of bounds' because they are situated in contested territory.[11]

On the other hand, there is growing evidence from studies in locations as varied as Jerusalem and Rio that improvements in public transportation can help bring about the sort of everyday casual encounters that help break down barriers between communities. In the case of Jerusalem, research has found that its light railway has made connections between parts of the city that until recent times had little public transport connectivity. This connectivity is enhanced by the railway's ability to link central commercial and transport hubs with peripheral neighbourhoods, helping to smooth out sharp economic divisions within the city (although, at times of tension, the increased connectivity can of course be problematic).[12] Mobility is equally important in integrating isolated communities in Latin American cities, where lower mobility affects patterns of encounter outside of an individual's social network. Differences in the ability to gain access to work and leisure and a lack of shared activity places seem to be the raw material conditions for social distancing and the installation of segregation in everyday life.[13]

A counter-example to the growing connectivity experienced in some cities lies in the lack of transport mobility that can be found in certain districts of Paris. The architect Léopold Lambert has demonstrated this phenomenon in his analysis of the peripheral neighbourhoods of Paris, the *banlieues*. In Figure 7.1 we can see one of his maps, coloured in shades from dark to light grey to represent a scale of low to high average

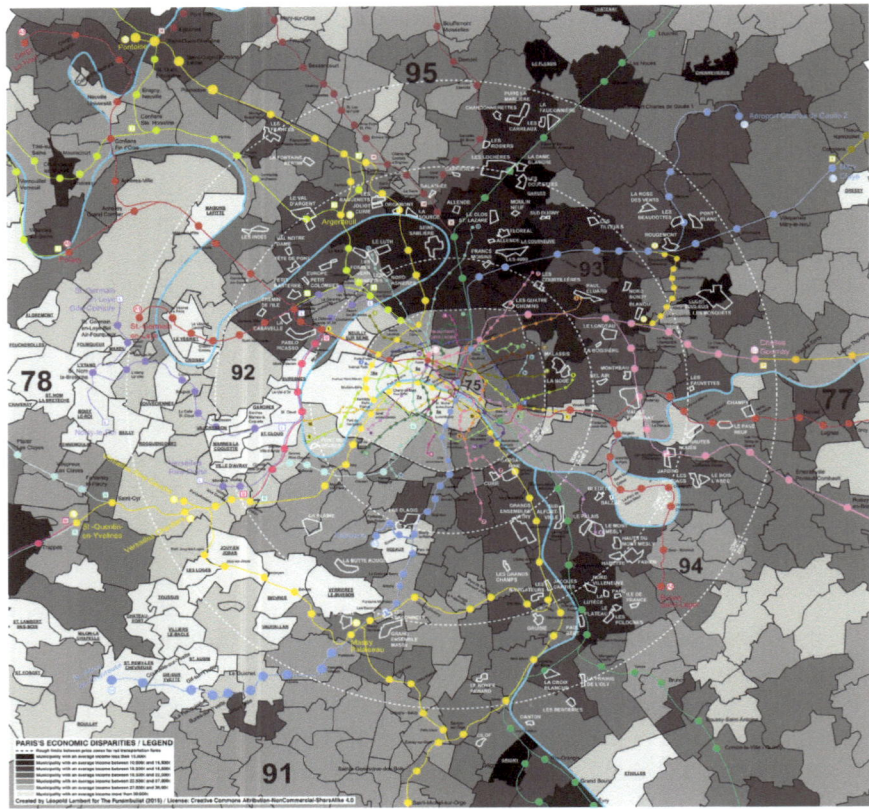

Figure 7.1 *Paris: wealth disparities*, 2015.
Created by Léopold Lambert for *The Funambulist* (2015). Copyright (CC BY-NC-SA 4.0).

income. It is overlaid with the location of *cités*, the post-war housing projects (marked as white outlines on the map).

The map shows how the spatial distribution of *cités* is very uneven across the city, with many located in areas of poverty in the remotest northeastern parts of the city. Constructed in concrete in a Brutalist style, the colossal, monolithic structures of the *cités* have become synonymous with crime, poverty, social isolation and high concentrations of the city's first- and second-generation citizens of North and West African origin.

A further refinement to this analysis can be found in another map produced by Lambert (Figure 7.2), which juxtaposes the location of the *cités* (coloured in red in this instance) with circles marking a 15-minute walking radius around each station within the *banlieues*. The pale grey

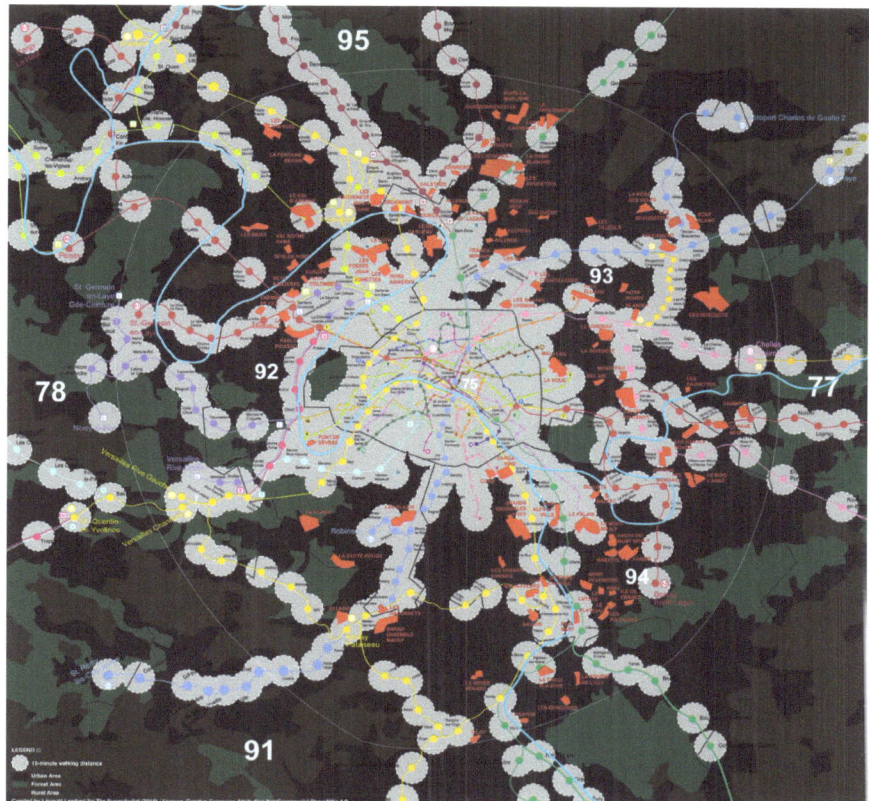

Figure 7.2 15-minute walking distance from a train station: spatial inequality in Paris *banlieues*.

Created by Léopold Lambert for The Funambulist (2015). Copyright (CC BY-NC-SA 4.0).

circles on the map indicate the maximum walking distance around each of the metro stations that can be covered in 15 minutes (the stations are marked by coloured dots, linked by coloured lines, which indicate the metro routes).[14] It shows that half the *cités* in the poverty districts are situated beyond the reach of the train stations. This means that if their inhabitants wish to get to work in central Paris, they are subject to lengthy, complicated journeys that will make it difficult for them to hold down a job. In fact, an average of 24 per cent of the population in the north and eastern districts of the periphery is unemployed.[15]

This mapping of a combination of social and transportation indicators highlights how the economic deprivation of a minority population can be exacerbated by a lack of access to a city's economic heartland.

Lambert writes elsewhere about the effect of the city's peripheral road on perceptions of the inhabitants of the Banlieue as being alienated from mainstream life:

> I have been repeatedly writing that the "boulevard périphérique" (highway ring) that surrounds Paris constitutes the contemporary equivalent of the city's fortress walls that used to be situated at the same place. The particularity of Paris is that the Parisian municipality only exists within these walls . . ., which has for consequence to substantially increase the centralised characteristics of the city in a country already far too centralized.[16]

While this centralisation of the city is of lesser importance to the wealthy districts to the west of the city, the inhabitants of the poorest districts of the city do not have the economic wherewithal to overcome their physical marginalisation from the city. Thus, while this is not a case of ghettoisation in the sense of the concentration of the bulk of a minority population in one location – the population of the *banlieues* is diverse ethnically, although marked by many more ethnic minority people than average – Paris's *banlieues* are an example of a problematic confluence of economic, ethnic and spatial segregation. Their predicament is worsened still by the specific history of France's minority Muslim population, who suffer from a legacy of discrimination that can make it harder for them to gain access to higher education and political power – though the tide is now turning on the latter to a certain extent.

In the case of the United States, the complexity of social geography is exemplified by many cities across the country, but the city of Chicago has taken centre stage in the past century and a half. This is due in no small part to the influential work of the Chicago School (though one might argue that the Chicago School's success came from its location within a definitive locus of urban problems). As we saw in Chapter 6, one of the School's most influential ideas emerged in its members' writings on the idea of the 'ghetto' in the 1920s and 1930s, when a model of immigrant integration was formulated as a three-stage progression from concentration to dispersal. Urbanism itself was being conceived as a temporal process driven by social heterogeneity. Yet Chicago was experiencing at that time some of the highest rates of black segregation in the country, with 83 per cent of the city's black population living in neighbourhoods that were themselves black-majority areas: at the extremely high rate of 93 per cent, the concentration of black residents in these areas was much higher than comparable clusters of other groups in their own respective areas, such as the Irish and Polish populations.[17] As Ceri Peach has pointed out, both historically and today, Chicago's black population in no way fits the Chicago School's tripartite model. This is true first for its scale, second in its extreme concentration as a single racial group and third in its lack of mobility over time: 'in the case of African Americans, outward movement did not always equate to dispersal. The ghetto moved out with them like the tongue of a glacier.'[18] Peach's analysis shows how the temporal continuity of racial division is at the heart of the problem of segregation. Not only this, when long-term racial division features alongside reduced access to the job market, socialisation, stigmatisation and limited access to social rights, the confluence of urban problems can be truly labelled as a ghetto. Indeed, features of this confluence were evident even earlier in the analysis of Philadelphia's black population by Du Bois.[19] We have also seen how the temporal longevity of urban segregation can be seen across the USA, where the past redlining of districts has led to dramatically different trajectories of wealth and deprivation within quite circumscribed areas, resulting in patterns of racial division on the ground that are not dissimilar to the situation in cities across South Africa, Latin America and the Middle East. The result is a drastic reduction in opportunities for everyday contact between people from different backgrounds.

The role of life on the street in shaping opportunities for social encounter has undoubtedly been a feature of many of the social maps considered in this volume, which have shown how the public houses of impoverished York or London provided a refuge from the degradations of life at home. They were also a place for local communities to cohere, for social life to be constituted and for economic activity to be formulated (whether legal or illegal). The street itself was a place for community interaction, with varying degrees of mixing across social groupings, though in the early years of this book's historical period this was more likely to be a working-class activity than one for the emerging middle classes, who could afford to remove themselves to the more refined arenas of the salon or (latterly) the department store. We should not though forget that use of the nineteenth-century city's streets was demarcated by sex as well as by class. Felicity Edholm has described how in late nineteenth-century Paris, 'working-class women spent most of their time within a quite tightly defined local area and would, within this area, be part of a community of women'.[20] She also describes how the working-class Parisienne had more freedom to explore her local streets than her bourgeoise counterpart, who could only venture out of the private sphere alone if it was to visit department stores; otherwise wealthier women experienced life on the street only if they were en route to cafés and restaurants with their husbands.

It is not a coincidence to find the café as central to social life throughout the nineteenth century. Scott Haine has highlighted the greater importance of the café in Paris than in most comparable cities in Europe or the United States. The working-class café provided a 'distinctive subculture' with its own rituals of behaviour and an ideal setting for creating 'an accessible, public, and open forum for social life'.[21] In other European cities such as Vienna and Warsaw, the café was central to the emergence of a political class, who used this quasi-public setting to formulate new ideas and in some cases as centres for assembly, organisation and political action.[22] Into the twentieth century, Michael Banton's study of black immigrants in London's East End found the café to be an important touchstone for people who had moved out of the area, to return on the weekend to meet their friends and to consume familiar food. In fact, his book included a map which recorded the location of a variety of cafés, distinguished by cultural background (African, Pakistani, Sikh or Mediterranean) and whether they served a predominantly immigrant clientele or catered primarily for local workmen.[23] As such, the role of the pub or the café (just as much as the role of the market, *suq* or bazaar) in public life is an often forgotten feature of the way in which marginalised people find a way on the one hand to negotiate a home away from home in the hostile city, and on the other to start to negotiate a position within society at large.[24] An interesting development of the latter has featured in recent work in South Africa, where Mpho Matsipa has found that a thoroughfare in Johannesburg, Bree Street, has become transformed by the arrival of new migrants with new modes of urban dwelling. She finds that the many different economic activities that are present on the street are transforming the way in which the area is being used.[25] The importance of minority ethnic businesses on prosaic streets for sustaining urban spatial justice is becoming apparent elsewhere in the world, from New York to Montreal, Amsterdam and London, where changes in cultural visibility can assist minority groups to assert their position in the public realm.

Visibility is important for reducing the mystery of difference, but when it is one-sided it can create an unbalanced set of power relations. Dana Katz describes how the Venice Ghetto was set up as a system for the

Figure 7.3 The Venice Ghetto and its houses of windows, with a synagogue atop the perpendicular heights of the buildings.

Photograph copyright Kayvan Karimi, 2016.

surveillance of its Jewish inhabitants 24 hours a day. The counterbalance to this was that the ghetto's buildings (which soared upwards, due to the limitations on spreading out at ground level) yielded 'extraordinary city sights' for those that lived at their heights. These could be contrasted with the constrained views of the Ghetto's guards (see Figure 7.3).[26]

These details of space and visibility allow us to view the ghetto as an extreme example of urban space in general. By studying the urban settings of social maps, we can start to grasp aspects of a social group as it is realised in its places of residence, its communal buildings and its places of commercial exchange and of religious worship.

Overall, the study of the street layout of social maps shows that if the street configuration is so contracted that it limits a minority's view on the world to a fractured, localised fragment, it can still function well to provide both local interactions and city-wide connections. This is dependent on the streets being sufficiently permeable to the area's courtyards, passages and streets, to ensure on the one hand that the world within it is not too mystified, that it is visible to the passer-by, and on the other hand that there are sufficient connections to the wider expanses of the city.

The politics of universal mapping

The preceding chapters have discussed nearly 50 maps – produced over a span of well over a century – each of which constitutes a social cartography of a singular time and place. Yet they also have in common an impetus to record a geography of an aspect (or many aspects) of society, alongside a drive to produce a graphic depiction that, if not scientific, is at least rigorous within its own terms. Nevertheless, we have seen a variety of ways in which visual rhetoric has been employed (whether implicitly or explicitly) to persuade, to campaign, to censure, or to define aspects of society that need improving in some way. Some aspects of social malaise have disappeared off the map, only to re-emerge, or to transmute into different materialisations of the same problems. We no longer have in cities such as London or New York the type of slums that drew the attention of Engels, Dickens or Jacob Riis, but such slums do continue to exist elsewhere in the world. Indeed, new formulations of slum living are taking shape at an astonishing pace, such that space standards set after the First World War for the United Kingdom are a long-distant dream in an era where so-called studio apartments are seen as fit for living for the twenty-first-century urban dweller. The same goes for disease: cholera is mostly eradicated in developed countries, but is an ongoing problem elsewhere in the world, periodically erupting along with natural disasters or man-made warfare. Tuberculosis has returned as a growing problem in many Western cities and continues to manifest wherever there are clusters of poverty and bad housing.

Disease, poverty, social and ethnic segregation – and indeed crime – are interconnected. Where one occurs, one or more of the other problems will follow. If this book had widened its scope it could have covered other issues still, such as maps indicating environmental inequality, transport barriers, political disenfranchisement and so on.[27] It is evident that the common factor in all of these is the spatial layout of the built environment. By studying maps along with the contemporaneous descriptions of their creators, it has become clear that the spatial isolation of a problem area can exert a powerful influence on urban social problems.

There was a relative dearth of social cartography in the latter half of the twentieth century. This may have been due to a change in focus in analysing urban problems but may also be due to a shift in authorship of maps. Whereas in this book's older maps the social map-maker was typically one or more individuals embedded in a locality, gathering data on an entire population and dependent on untutored or self-tutored statistical expertise, by the early twentieth century authorship had shifted to a professional, university-based milieu. Data-gathering was still undertaken in the field, but analysis was more remote, both physically and conceptually. No longer did we see accounts by people who knew each of the data points on the map as living, breathing individuals. This may have been advantageous – with distance comes perspective, yet also a loss of clarity. In parallel to this, the increase in professionalisation of social research meant that the size of social maps could increase concomitantly. Ultimately, social cartography shifted from studying complete, discrete areas to sampling methods that enabled researchers to analyse entire countries and indeed to compare nations.[28] Despite the decline in social maps in the later decades of the twentieth century, social cartography is being increasingly used in a wide variety of governmental and research arenas as well as by the general public. This has stemmed from two important technological changes.

The first change has been the exponential increase in computing capacity, which has meant that governmental bodies from the local to the national can capture an ever-increasing number of data points on their citizens, from crime incidents to place of birth, annual income and patterns of consumption. The second change has been the development of geospatial systems such as Google Maps, with the outcome that anyone with access to a personal computer has at their fingertips an unimaginable amount of information on location, presented in a graphic format that is comprehensible much more readily than the statistical tables of figures that governments have published in the past.

This apparent democratisation of spatial data is at first glance a positive move, yet in many ways this is an incorrect perception. Who owns the data, as well as who owns the means of gathering and presenting the data, will shape how it is presented (and what is not presented). The many issues with distortions of data representation and the problems with interpretation that were apparent in the historical maps are frequently present in contemporary maps too, such as the choice of where to segment the data when drawing up categories or where to draw the boundary around an area in order to determine its social character. In the latter case, the picture of the spatial location of poverty clusters in a city may be distorted by the tendency of poverty to be concentrated in small areas. Similarly, burglary rates might look dramatically different according to where an area's boundary line is drawn.

Mapping specialists will be aware of these problems; they are used to taking account of the spatial variation of data, problems of scale and aggregation and moreover the ecological fallacy already mentioned. Yet less tutored map users may very well gain false impressions of the state of their neighbourhood or city.

Moreover, while in the past governments have produced representations of their cities in order to orientate or control their populations, nowadays the accessibility of geo-information on society gives the misleading impression that such representations are objective, definitive pictures of a given place in time. This is the case not only because of the seductive amplitude of maps, but also because they give the impression that they are comprehensive and that they are essentially truthful. Tools such as GPS (Global Positioning Systems) are typically presumed to be neutral, and their usefulness for humanitarian activities is undoubtedly laudable, but the same tools are employed for the surveillance of individual citizens by governments and by third parties. The dominance of only a handful of mapping systems that are held in the hands of a few companies is an additional problem. There is no simple answer to the conundrum of the balance to be found between the power and utility of mapping systems and issues of personal privacy or loss of power, although the emergence of citizen scientists who gather data on (and in) their own backyards, whether to hold organisations to account or to provide free alternatives to corporate mapping agencies, makes for a possible redress to this imbalance.[29] It is interesting to note in this closing section a recent paper which argues that citizen science constitutes a reversion to the nineteenth-century non-professional participation in scientific endeavours, when many local societies participated in public health campaigns against environmental pollution.[30]

However, this book does not intend to focus on a critique of mapping methods, whether in the past or in the present, but to highlight the significance of spatial, geographical factors that are revealed by maps, even taking account of the limitations of social cartography. Importantly, one of the most constant features of the historical maps presented in this book has been their demonstration of the importance of urban form and spatial configuration in shaping social outcomes: this is repeatedly substantiated by the fact that urban problems frequently persist over many years, even decades. How social cartography will develop in the future, whether it will become the domain of government, large corporations or individual citizen scientists, is a story still to be told.

Notes

1. N. Hildyard, 'Building for Collapse,' *Ecologist* 7, no. 2 (1977), p. 53.
2. More recently the United Nation's Sustainable Development Goals has updated this to 17 goals and 169 indicators.
3. B. Hillier and L. Vaughan, 'The City as One Thing,' *Progress in Planning: special issue on The Syntax of Segregation, edited by Laura Vaughan* 67, no. 3 (2007), p. 205.
4. L. Wacquant, *Urban Outcasts: A Comparative Sociology of Advanced Marginality* (Cambridge: Polity, 2007). S. Sassen, 'The Global City: The De-Nationalizing of Time and Space' (paper presented at the conference El territori en las societat de les xarxes. Dinámiques territorials i organització territorial, Barcelona, 2–3 October 2000).
5. Dwyer, 'Poverty, Prosperity, and Place'; G. MacLeod and K. Ward, 'Spaces of Utopia and Dystopia: Landscaping the Contemporary City,' *Geografiska Annaler: Series B, Human Geography* 84, no. 3–4 (2002).
6. E.W. Soja, *Postmodern Geographies: Reassertion of Space in Critical Social Theory* (London: Verso, 1989). *Banlieue* is the term for Paris's peripheral suburbs. The term 'banlieue' is by no means neutral. It has come to describe an especially bleak, disordered, landscape, both physically and conceptually.
7. P. Marcuse, 'Putting Space in Its Place: Reassessing the Spatiality of the Ghetto and Advanced Marginality,' *City* 11, no. 3 (2007), p. 380.
8. R. Sennett, *The Foreigner* (London: Notting Hill Editions, 2011), 24.
9. Northall, *Travels through Italy*, p. 128.
10. G. Simmel, 'The Stranger,' in *The Sociology of Georg Simmel*, ed. K.H. Wolff (New York: Free Press, 1950); first published in German as 'Exkurs über den Fremden,' in *Soziologie: Untersuchungen über die Formen der Vergesellschaftung* (Leipzig: Duncker & Humblot, 1908).
11. B. Murtagh and P. Shirlow, 'Spatial Segregation and Labour Market Processes in Belfast,' *Policy & Politics* 35, no. 3 (2007). In this context, maps are increasingly being used to good effect to change public reading of areas of avoidance, with the Belfast Interface Project recording the historical construction of peace lines as well as progress in their removal https://www.belfastinterfaceproject.org/interfaces-map.
12. J. Rokem and L. Vaughan, 'Segregation, Mobility and Encounters in Jerusalem: The Role of Public Transport Infrastructure in Connecting the "Divided City",' *Urban Studies* (2017).
13. V. M. Netto, M. S. Pinheiro, and R. Paschoalino, 'Segregated Networks in the City,' *International Journal of Urban and Regional Research* 39 (2015), p. 1099.
14. The calculation is only a rough indicator, as it does not take account of where people might be able to walk. A 15-minute walk along the streets is likely not to reach as far as the full radius, especially within the vicinity of a railway station, which normally creates interruptions to the continuity of the street network.
15. A map of unemployment produced by an EU study of inequalities shows a range of 13–26 per cent unemployment in the outskirts of Paris, particularly in the northern and eastern districts, and in a few clusters in the south. INEQ-CITIES, 'Paris: Socio-Economic Indicators,' UCL, https://www.ucl.ac.uk/silva/ineqcities/atlas/cities/paris/paris-sei#prim. Accessed 10 March 2018.
16. L. Lambert, 'Another Paris: The Banlieue Imaginary,' *The Funambulist Papers*, https://thefunambulist.net/architectural-projects/maps-another-paris-the-banlieue-imaginary. Accessed 7 September 2017.
17. T.L. Philpott, *The Slum and the Ghetto: Neighborhood Deterioration and Middle-class Reform, Chicago, 1880–1930* (New York: Oxford University Press, 1978), cited in C. Peach, 'The Ghetto and the Ethnic Enclave,' in *Desegregating the City: Ghettos, Enclaves and Inequality*, ed. D.P. Varady (Albany: State University of New York Press, 2005).
18. Peach, 'The Ghetto and the Ethnic Enclave', p. 37.
19. Du Bois, *The Philadelphia Negro*.
20. Edholm, 'The View from Below,' pp. 159–60.
21. W.S.C. Haine, *The World of the Paris Cafe: Sociability among the French Working Class, 1789-1914* (Baltimore, MD, and London: Johns Hopkins University Press, 1996), p. ix.
22. Scott Ury has shown how bourgeois society, secular culture and Jewish politics came together in Warsaw's turn-of-the-twentieth-century cafés, providing 'an open, public space where patrons could meet to discuss cultural affairs, promote communal projects and organize political activities beyond the watchful eye of traditional community leaders and Tsarist officials'.

S. Ury, 'Common Grounds? On the Place and Role of Jewish Coffee Houses at the Turn of the Century,' paper presented at a conference on Warsaw – the History of a Jewish Metropolis, 22–25 June 2010, University College London.
23. Banton, *The Coloured Quarter.*
24. L. Vaughan, 'The Ethnic Marketplace as Point of Transition,' in *London the Promised Land Revisited*, ed. A. Kershen, Studies in Migration and Diaspora (Farnham, Surrey: Ashgate, 2015).
25. M. Matsipa, 'Street Values: Johannesburg's Spatial Agents,' *Architectural Review*, 4 July 2016.
26. The immense height of the ghetto's buildings meant that it was visible from a distance, despite its position on the margins. D. Katz, '"Clamber Not You Up to the Casements": On Ghetto Views and Viewing,' *Jewish History* 24, no. 2 (2010), p. 137; D.E. Katz, *The Jewish Ghetto and the Visual Imagination of Early Modern Venice* (Cambridge: Cambridge University Press, 2017), p. 52. This analysis of urban visibility at the finest scale of buildings, courtyards and streets is rarely considered in the context of studying social maps.
27. In the case of political analysis of urban space, the work of Tobias Metzler on European Jewry between the wars is of interest. His analysis shows the relationship between the locations and the extents of the spatial distribution of Jewish residential clusters and the location of polling stations for the community's internal elections in Berlin, 1926. T. Metzler, *Tales of Three Cities: Urban Jewish Cultures in London, Berlin, and Paris (1880–1940)*, vol. 028, Jüdische Kultur (Wiesbaden: Harrassowitz, 2014), p. 217. Similarly, Sadaf Sultan Khan's analysis of Muhajir minority electoral patterns in Karachi compares political voting patterns to the location of the minority group's most commonly frequented mosques to explore the relationship between place and politics. S.S. Khan, K. Karimi and L. Vaughan, 'The Tale of Ethno-Political and Spatial Claims in a Contested City: The Muhajir Community in Karachi,' in *Urban Geopolitics: Rethinking Planning in Contested Cities*, ed. J. Rokem and C. Boano, Routledge Studies in Urbanism and the City (Abingdon: Routledge, 2017).
28. In parallel, social anthropology has grown as a field of expertise, filling the gap left by the early social scientists by studying single communities in depth.
29. Examples of citizen science projects include *Zooniverse*, which has enabled the public to contribute to research by active participation in gathering data, classifying it and even processing it. A variant on standard citizen science argues for reciprocity between the public and the scientist, so that the former benefits from the scientific findings of the research, such as a project that trains non-literate forest communities in the Congo Basin to use handheld GIS devices for mapping their environment. See more about this in M. Vitos et al., 'Supporting Collaboration with Non-Literate Forest Communities in the Congo-Basin,' in *CSCW '17: Proceedings of the 2017 ACM Conference on Computer Supported Cooperative Work and Social Computing* (New York: ACM, 2017). Other examples include *OpenStreetMap* and the *Missing Maps* project – a humanitarian project that pre-emptively maps part of the world vulnerable to natural disasters, conflicts and disease epidemics: http://www.missingmaps.org/.
30. S. Shuttleworth and S. Frampton, 'Constructing Scientific Communities: Citizen Science,' *The Lancet* 385, no. 9987 (2015). Arguably the Temperance Societies were like this, in the sense that they were driven by individuals and local groups more than the organised church.

Appendix
The spatial syntax of society

Ideas regarding the morphology of social space are foundational to space syntax theory, which argues that society does more than simply exist in space: it also takes on a definite spatial form, arranging people in space, locating them in relation to each other. People do not just inhabit space; they also shift their position in their locality throughout the day and the week and over longer periods of time. As cities become more diverse socially, they become correspondingly more likely to have a complex arrangement of streets, buildings, zones or neighbourhoods, which means that at first glance any social pattern that appears in cities seems to be unconnected to its setting. Countering this, space syntax theory argues that urban space has its own formal logic that is shaped by the way in which society is organised – not in a one-to-one relationship, but as the result of a circular process of society shaping urban space and urban space then giving rise in turn to certain regularities in social behaviour. Whilst scholars of urban sociology such as Herbert Gans recognise that social systems adapt over time (see his classic, *The Levittowners*, for example),[1] they will commonly argue that social space is merely a container of society, that it will not influence behaviour within it. Indeed, to propose that there is any deterministic relationship between space and society would be anathema to some social scientists.[2] By contrast, this book argues that the physical milieu of society takes on a definite shape in response to social patterns.[3]

The ability of the city to form spatial patterns is the means through which it is also able to give expression to social meanings. In this way, space syntax theory and methodology provide a fundamental description of social space, using the basic elements of social space: the line, the convex space and the isovist, each of which represents an aspect of how human beings inhabit physical space (see Figure a): people move through space, interacting with other people in space, or even just seeing ambient space from a point in it, and each of these actions has a natural and necessary spatial geometry. Movement is essentially linear (see blue lines in Figure a), interaction requires a *convex* space in which all points can see all others (see green quadrilateral), and from any point in space a person will see a variably shaped, often spiky, visual field (see orange shape, the isovist, in Figure a).

Each of these geometric elements captures some aspect of how we use or experience space, and assembling them into a spatial model allows us to see how buildings and cities are organised in terms of these geometric ideas. In fact, urban space will always have all three properties, and by being clear about the nature of this geometry we can begin to systematically describe and analyse cities as social systems.

The second fundamental concept in space syntax theory is that human space is defined both by the properties of individual spaces and by the inter-relations between the many spaces that make up the spatial layout of a building or a city. Taken together, space syntax methods provide the tools for a mathematical analysis of these inter-relations. In other words, the street network is analysed as a spatial configuration.

The analysis first theorises the spatial configuration in terms of its potential to embody or transmit social ideas, and then turns it into measures and representations of spatial structure by linking spatial measures to social measures (such as land use patterns or burglary cases).[4] For the purpose of urban analysis, this will involve constructing an axial map, which is a simplified representation of all the public open space as a network of lines (see blue lines in Figure a), using basic measures of graph mathematics starting with connectivity: how many times

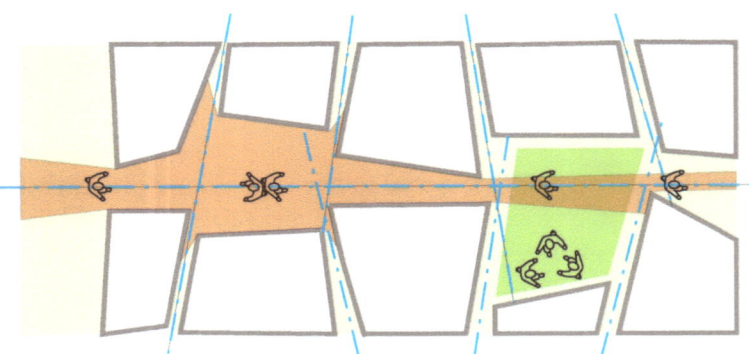

Figure a   Space is not a background to activity, but an intrinsic aspect of it.

each street is connected to its neighbours, and so on. More complex measures follow suit: calculating the relative connectivity of each space to all other spaces within a given distance, taking account of angular change, and many additional variants. In their essence, the basic space syntax measures are formal interpretations of the notion of spatial integration and segregation, and it was the formalisation of these terms during early space syntax studies of myriad settlements around the world in the 1970s which resulted in fundamental findings regarding the social logic of spatial systems.[5] The results of this research provided the tools for archaeologists to formulate hypotheses about the social characteristics of prehistoric settlements and for anthropologists to investigate the social nature of contemporary society.[6] For urban design and planning specialists, the provision of a measurable scale of social space from *segregation* to *integration* enabled statistical comparison of different spatial forms across cultures, and so provided a platform from which the social origins and consequences of urban settlements might be investigated.[7]

This was an unfamiliar idea. Early space syntax theory was, in effect, suggesting that space has its own formal logic prior to acquiring a social logic, and indeed that it was this logic of space that was exploited to render space *social*. This is the core argument of 'The Social Logic of Space', the first book-length text published in the field.[8] The autonomous potential of space to form social patterns was, in effect, seen as the means through which it can give expression to social meanings. These ideas challenged paradigms in both the social and the spatial sciences. But they also suggested that there could be an approach to urban research which was both quantitative and at the same time could be informed by investigations into social and cultural influences and meanings.

From these beginnings, space syntax has evolved into a set of tools linked to a set of theories, the two together giving rise to a set of interpretative models for different socio-spatial phenomena. Interpretative models are schemes of analysis which work for specific phenomena. Space syntax analysis of maps and plans produces both visual and numerical descriptive data which can be used to formulate hypotheses – for example about movement and encounter patterns in the past – that can help in interpreting the significance of non-cartographic historical sources. This is especially useful when historical sources provide data at street or building lot scale (for example census or business directory data), since then the delineation of urban activities, such as the diversity of business types or the structure of a kinship network, can be read back from the plan as a socio-spatial as well as a socio-economic artefact. An early example of the application of space syntax in historical research can be found in a study of the Charles Booth maps of poverty, which follows below.

Because space syntax analysis is concerned with systematically describing and analysing streets configurationally, it represents the city's public streets and squares as a continuous system, preparing the ground for analysis of how well connected each street space is to its surroundings. This is done by taking an accurate map and drawing a set of intersecting lines through all the spaces of the urban grid so that the grid is covered, and all rings of circulation are completed. The resulting set of lines is called an 'axial map': an example can be seen in Figure b, which shows the Booth map of poverty in London, 1889, overlaid with the axial map (the black lines) of the map's street layout.

Space syntax analysis computes all the lines in the axial map according to their relative depth to each other. Segment maps enable us to represent continuous open space at a finer scale. They are generated by breaking axial lines into segments at the intersections of the axial lines. They allow for differences between each street segment to be calculated. The results from such mathematical calculations are used to index the maps in a range from red (high) to blue (low) to represent the distribution of spatial values, allowing the researcher to visually inspect the map for observable patterns. We can see the results of one of

Figure b   Axial line map of the East End district of London, overlaid on Booth's 1889 map.

Figure c Detail of *Descriptive Map of London Poverty* 1889, showing the East End district of London, overlaid with space syntax analysis of spatial accessibility for each street to all other streets within 800m.[9]

the analyses in Figure c, which is coloured up according to integration radius 800, namely a measure of how accessible each segment is from its surroundings, limiting the computation to 800m measured along the street. The image illustrates how the main streets of the area in 1889 were considerably more accessible than the back streets. Notably, this analysis is conducted independently of the data on poverty, so the two factors can be compared statistically.

While the visual patterns are useful, they cannot substantiate research findings, so the data on each street's spatial values are related in a statistical table to social data. In this instance it is levels of poverty, but other studies will analyse the location and type of crime, or the diversity of land uses. By examining space in this way, we can analyse the correspondence between spatial segregation/integration and social statistics (namely the precise data underlying the maps shown in Figures d and e). For example, we can research whether there is a relationship between the locations of burglaries and housing layout; or, whether more successful shopping streets have spatial characteristics in common. In the case of the Booth map of poverty, while the maps reveal that streets with greater poverty are frequently locations with less physical access, this can only be substantiated empirically by plotting the results in a statistical graph. The full analysis of poverty and spatial segregation was published in a series of papers and is summarised in Chapter 3, but it is noteworthy that the study found a strong correspondence existed between all the poverty scales and spatial segregation. In other words, Booth's reading of a relationship between the location of poverty and the physical pattern of the city could (as has been elucidated in detail by Harold Pfautz) be substantiated empirically.[10]

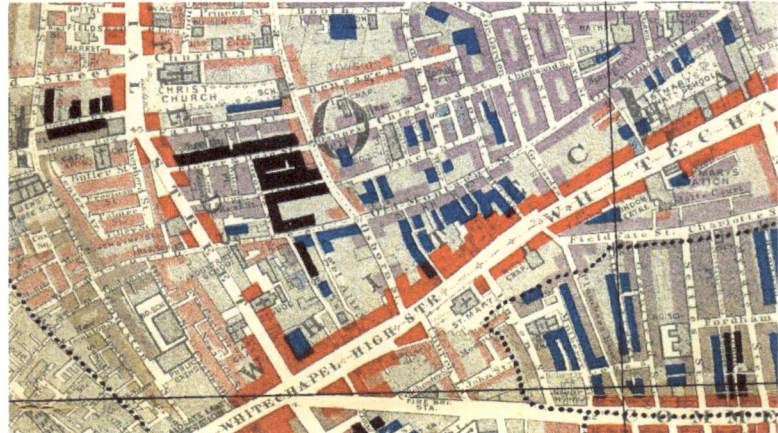

Figure d Detail of *Descriptive Map of London Poverty* 1889, showing the East End district of London.

Figure e Detail of space syntax analysis of spatial accessibility for each street to all other streets within 800m.[11]

Notes

1. H. Gans, *The Levittowners* (New York: Pantheon Books, 1967).
2. Both in his earliest work (H. Gans, 'Planning and Social Life: Friendship and Neighbour Relations in Suburban Communities,' *Journal of the American Institute of Planners* 27 (1966)) and more recently (in H. Gans, 'Some Problems of and Futures for Urban Sociology: Toward a Sociology of Settlements,' *City & Community* 8, no. 3 (2009)), Gans has continued to argue that social space is a mere container of society, though he notes that 'one of the dominant forms of spatial analysis has a very different agenda, to show that space and place have independent social effects that can shape a variety of aspects of social life prior to social intervention. . . Such 'physical' determinism may facilitate architectural thought but it is not a sociological analysis' (Gans, 'Some Problems of and for Urban Sociology,' p. 216).
3. As was first laid out in Hillier and Hanson's *The Social Logic of Space*, p. 27.
4. See further discussion in the introduction to L. Vaughan, 'The Spatial Syntax of Urban Segregation,' *Progress in Planning* 67, no. 3 (2007).
5. B. Hillier et al., 'Creating Life: Or, Does Architecture Determine Anything?' *Architecture & Behaviour* 3, no. 3 (1987).
6. See for example J. Shapiro, *A Space Syntax Analysis of Arroyo Hondo Pueblo, New Mexico: Community Formation in the Northern Rio Grande*, vol. 9, Arroyo Hondo Archaeological Series (Santa Fe, New Mexico: School of American Research Press, 2005).
7. As well as urban and settlement analysis, the method has been shown to be equally useful in explaining cultural variations in domestic layouts and the internal configuration of complex buildings such as museums and hospitals.
8. Hillier and Hanson, *The Social Logic of Space*.
9. The measure used was segment angular integration radius 800. Segment angular integration measures how close each segment is to all others in terms of the sum of angular changes that are made on each route.
10. Pfautz, *On the City*; Vaughan and Geddes, 'Urban Form and Deprivation.'
11. See note to Figure c.

References

Acheson, D. 'Report of the Independent Inquiry into Inequalities in Health.' London: Stationery Office; Department of Health, 1998.

Acland, H.W. *Memoir on the Cholera at Oxford, in the Year 1854, with Considerations Suggested by the Epidemic*. London: John Churchill, 1856.

Alexander, S. 'A New Civilization? London Surveyed 1928–1940s.' *History Workshop Journal* 64, no. 1 (21 September 2007): 296–320.

Allen, R. 'Flint Map: Where Lead Levels in Water Remain Too High.' *Detroit Free Press*, 2 February 2016. http://www.freep.com/story/news/local/michigan/flint-water-crisis/2016/02/02/flint-lead-map/79686158/.

Anonymous. 'Jewish East London.' *The Jewish Chronicle*, 26 April 1889, 9.

Atkins, P.J. 'The Spatial Configuration of Class Solidarity in London's West End 1792-1939.' *Urban History* 17, no. -1 (1990): 36–65.

Atkinson, J.F. 'A Pathetic Case.' In *The Missionary Visitor*, edited by Brethren's General Missionary and Tract Committee, 46–7. Elgin, Illinois: General Missionary and Tract Committee, 1905.

Bailey, R., and A. Leith. *Computerising and Coding the New Survey of London Life and Labour*. Colchester: University of Essex, 1997.

Balch, E.G. 'Hull House Maps and Papers: Review of A Presentation of Nationalities and Wages in a Congested District of Chicago, Together with Comments and Essays on Problems Growing Out of the Social Conditions.' *Publications of the American Statistical Association* 4, no. 30 (1895): 201–3.

Bales, K. 'Charles Booth's Survey of Life and Labour of the People in London 1889–1903.' In *The Social Survey in Historical Perspective, 1880-1940*, edited by M. Bulmer, K. Bales and K. Kish Sklar, 66–110. Cambridge: Cambridge University Press, 1991.

Bales, K. 'Popular Reactions to Sociological Research: The Case of Charles Booth.' *Sociology* 33, no. 01 (1999): 153–68.

Ball, R., and J. Drury. 'Representing the Riots: The (Mis)Use of Figures to Sustain Ideological Explanation.' *Radical Statistics*, no. 106 (2012): 4–21.

Banton, M. *The Coloured Quarter: Negro Immigrants in an English City*. London: Cape, 1955.

Barford, A., and D. Dorling. 'Mapping Disease Patterns.' In *Wiley StatsRef: Statistics Reference Online*, John Wiley & Sons, Ltd, 2014.

Baudains, P., A. Braithwaite and S.D. Johnson. 'Target Choice During Extreme Events: A Discrete Spatial Choice Model of the 2011 London Riots.' *Criminology* 51, no. 2 (2013): 251–85.

Beirne, P. *Inventing Criminology: Essays on the Rise of 'Homo Criminalis'*. Albany: State University of New York Press, 1993.

Bennett, P. 'Geographies of Financial Risk and Exclusion.' In *The Sage Handbook of Social Geographies*, edited by S.J. Smith, R. Pain, S.A. Marston and J.P. Jones, 222–36: SAGE Publications, 2009.

Blair, H.W. *The Temperance Movement, or the Conflict between Man & Alcohol*. Boston: William E. Smyth Co., 1888.

Blake, A.M. *How New York Became American, 1890–1924*. Baltimore: Johns Hopkins University Press, 2009.

Booth, C. 'Poverty Series Survey Notebooks (Online Archive).' British Library of Political and Economic Science, https://booth.lse.ac.uk/notebooks/.

Booth, C. 'The Inhabitants of Tower Hamlets (School Board Division), Their Condition and Occupations.' *Journal of the Royal Statistical Society* 50, no. 2 (1887/06 1887): 326–401.

Booth, C. 'Condition and Occupations of the People of East London and Hackney, 1887.' *Journal of the Royal Statistical Society* 51, no. 2 (1888/06 1888): 276–339.

Booth, C. *Labour and Life of the People*. Appendix to volume II, ed. Charles Booth. London and Edinburgh: William and Norgate, 1891.

Booth, C. 'Poor Law Statistics.' *The Economic Journal* 6, no. 21 (1896/03 1896): 70–4.

Booth, C. *Old Age Pensions and the Aged Poor: A Proposal*. London: Macmillan and Company, 1899.

Booth, C. *Improved Means of Locomotion as a First Step Towards the Cure of the Housing Difficulties of London*. Abstract of the Proceedings of Two Conferences Convened by Albert Browning Hall, Walworth. London: Macmillan, 1901.

Booth, C. *Life and Labour of the People in London*. 3rd series. 17 vols. London: Macmillan and Co., 1903.

Booth, C. *Life and Labour of the People of London*. Vol. 1: East, Central and South London. London: Macmillan and Co., 1904.

Booth, W. *In Darkest England and the Way Out*. New York and London: Funk & Wagnal, 1890.

Borchert, J. *Alley Life in Washington: Family, Community, Religion, and Folklife in the City, 1850-1970*. Urbana: University of Illinois Press, 1980.

Boston (Mass.) Committee on Internal Health. Report of the Committee of Internal Health on the Asiatic cholera, together with a report of the city physician on the Cholera Hospital. Boston, MA: J.H. Eastburn, 1849.

Brantingham, P.J., and P.L. Brantingham. *Environmental Criminology*. Beverly Hills and London: Sage Publications, 1981.

Bray, W. ed. *Memoirs of John Evelyn ... Comprising His Diary, from 1641 to 1705-6, and a Selection of His Familiar Letters*. London: Henry Colburn, 1827.

Breward, C. 'Fashion's Front and Back: "Rag Trade" Cultures and Cultures of Consumption in Post-War London c.1945-1970.' *The London Journal* 31, no. 1 (2006, June): 15–41.

Bullman, J., N. Hegarty and B. Hill. *The Secret History of Our Streets - London: A Social History through the Houses and Streets We Live In*. London: BBC Books and Random House, 2012.

Bulmer, M., K. Bales and K. Kish Sklar, eds. *The Social Survey in Historical Perspective, 1880-1940*. Cambridge: Cambridge University Press, 1991.

Buman, M.P., E.B. Hekler, W.L. Haskell, L. Pruitt, T.L. Conway, K.L. Cain, J.F. Sallis, et al. 'Objective Light-Intensity Physical Activity Associations with Rated Health in Older Adults.' *American Journal of Epidemiology* 172, no. 10 (2010): 1155–65.

Bunge, W. *Nuclear War Atlas*. Oxford: Basil Blackwell, 1988.

Bureau of Associated Charities, Newark, NJ, and A.W. MacDougall. *The Resources for Social Service, Charitable, Civic, Educational, Religious, of Newark, New Jersey; a Classified and Descriptive Directory*. New York: G. P. Putnam's Sons, 1912.

Burgess, E.W. 'The Growth of the City: An Introduction to a Research Project.' In *The City*, edited by R.E. Park, E.W. Burgess and R.D. McKenzie, 47–62. Chicago University of Chicago Press, 1925.

Calabi, D. *Venice and Its Jews: 500 Years since the Founding of the Ghetto (Translated by Leonore Rosenberg)*. Milan: Officina Libraria, 2017.

Campkin, B. *Remaking London: Decline and Regeneration in Urban Culture*. Vol. 19. London and New York: IB Tauris, 2013.

Cannadine, D., and D. Reeder, eds. *Exploring the Urban Past: Essays in Urban History by H. J. Dyos*. Cambridge: Cambridge University Press, 1982.

Canoui-Poitrine, F., E. Cadot and A. Spira. 'Excess Deaths During the August 2003 Heat Wave in Paris, France.' *Revue d'Épidémiologie et de Santé Publique* 54, no. 2 (Apr 2006): 127–35.

Chadwick, E. 'Report on the Sanitary Conditions of the Labouring Population of Great Britain.' In *Report to Her Majesty's Principal Secretary of State for the Home*

Department, from the Poor Law Commissioners, on an Inquiry into the Sanitary Condition of the Labouring Population of Great Britain. London: W. Clowes and Sons, 1842.

Chung, S. 'London Chinatown: An Urban Artifice or Authentic Chinese Enclave?' Paper presented at the City Street International Conference – First International Conference on City Street, Cultural Intimacy and Global Image, Notre Dame University, Louaize, Lebanon, 18–20 November 2009.

Cicak, T., and N. Tynan. 'Mapping London's Water Companies and Cholera Deaths.' *The London Journal* 40, no. 1 (2015, March): 21–32.

Clarke, B. 'Mapping the Methodologies of Burkitt Lymphoma.' *Studies in History and Philosophy of Science Part C: Studies in History and Philosophy of Biological and Biomedical Sciences* 48 (2014, December): 210–17.

Collins, D. 'The Introduction of Old Age Pensions in Great Britain.' *The Historical Journal* 8, no. 2 (1965): 246–59.

Cosgrove, D. *Geography and Vision: Seeing, Imagining and Representing the World*. London and New York: I.B. Tauris, 2012.

Cottret, B. *The Huguenots in England: Immigration and Settlement 1550–1700*. Cambridge: Cambridge University Press, 1991.

Couling, S. *History of the Temperance Movement in Great Britain and Ireland*. London: W. Tweedie, 1862.

Croxford, B., A. Penn and B. Hillier. 'Spatial Distribution of Urban Pollution: Civilizing Urban Traffic.' *Science of The Total Environment* 189 (28 October 1996): 3–9.

Cutler, K.-M. 'East of Palo Alto's Eden: Race and the Formation of Silicon Valley.' *TechCrunch.com* (2015, January 10): http://techcrunch.com/2015/01/10/east-of-palo-altos-eden/.

Daudé, É., E. Eliot and E. Bonnet. 'Cholera in the 19th Century: Constructing Epidemiological Risk with Complexity Methodologies.' Paper presented at the Third International Conference on Complex Systems and Applications, University of Le Havre, Normandy, France, 29 June to 2 July 2009.

Davin, A. *Growing up Poor: Home, School and Street in London 1870–1914*. London: Rivers Oram Press, 1996.

Dearle, N.B. 'Review: The New Survey of London Life and Labour. Volumes III and IV: Survey of Social Conditions. I. Eastern Area and Eastern Area Maps.' *The Economic Journal* 43, no. 170 (1933): 315–19.

Denison, E., M. Teklemariam and D. Abraha. 'Asmara: Africa's Modernist City (UNESCO World Heritage Nomination).' *The Journal of Architecture* 22, no. 1 (2017/01/02 2017): 11–53.

Dennis, R. *Cities in Modernity: Representations and Productions of Metropolitan Space, 1840-1930*. Cambridge: Cambridge University Press, 2008.

Dhanani, A., Tarkhanyan, L. and Vaughan, L. 'Estimating Pedestrian Demand for Active Transport Evaluation and Planning.' *Transportation Research Part A: Policy and Practice* 103 (2017), 54–69.

Dickens, C. *Nicholas Nickleby*. University of Oxford Text Archive http://ota.ox.ac.uk/text/3082.html. Accessed 24 May 2017.

Dickens, C. *Sketches by Boz - Complete in One Volume*. Paris: Baudry's European Library, 1839.

Dorling, D. 'Housing and Identity: How Place Makes Race.' *Better Housing Briefing Paper 17*. London: Race Equality Foundation, 2011.

Dorling, D., R. Mitchell, M. Shaw, S. Orford and G. Smith. 'The Ghost of Christmas Past: Health Effects of Poverty in London in 1896 and 1991.' *British Medical Journal* 321 (23–30 December 2000): 1547–51.

Douglas, M. *Purity and Danger: An Analysis of Concepts of Pollution and Taboo*. London: Routledge and Kegan Paul, 1966.

Dousset, B., F. Gourmelon, K. Laaidi, A. Zeghnoun, E. Giraudet, P. Bretin, E. Mauri, and S. Vandentorren. 'Satellite Monitoring of Summer Heat Waves in the Paris Metropolitan Area.' *International Journal of Climatology* 31, no. 2 (2011): 313–23.

Driver, F. 'Moral Geographies: Social Science and the Urban Environment in Mid-Nineteenth Century England.' *Transactions of the Institute of British Geographers* 13, no. 3 (1988): 275–87.

Du Bois, W.E.B. *The Philadelphia Negro: A Social Study*. New York: Schocken Books, 1899.
Du Bois, W.E.B. *The Souls of Black Folk*. Chicago: McClurg & Co., 1903.
Du Bois, W.E.B. 'My Evolving Program for Negro Freedom.' In *What the Negro Wants*, edited by R.W. Logan. Indiana: University of Notre Dame Press, 1944.
Duneier, M. *Ghetto: The Invention of a Place, the History of an Idea*. New York: Farrar, Straus & Giroux, 2016.
Durkheim, E. *The Division of Labour in Society (English Edition, 1964)*. Translated by G. Simpson. English ed. New York: Macmillan Publishing Company, 1893.
Dwyer, C., D. Gilbert and B. Shah. 'Faith and Suburbia: Secularisation, Modernity and the Changing Geographies of Religion in London's Suburbs.' *Transactions of the Institute of British Geographers* 38, no. 3 (2013): 403–19.
Dwyer, R.E. 'Poverty, Prosperity, and Place: The Shape of Class Segregation in the Age of Extremes.' *Social Problems* 57, no. 1 (2010): 114–37.
Dyos, H.J. 'The Slums of Victorian London.' *Victorian Studies* XI (1967): 5–40.
Dyos, H.J., and M. Wolff, eds. *The Victorian City: Images and Realities (Past and Present & Numbers of People)*. 2 vols. Vol. 1. London, Henley and Boston: Routledge & Kegan Paul, 1976.
Edholm, F. 'The View from Below: Paris in the 1880s.' In *Landscape: Politics and Perspectives*, edited by B. Bender. Explorations in Anthropology: A University College London Series, 139–68. Oxford: Berg Publishers, 1995.
Engels, F. *The Condition of the Working Class in England*. New York and London: Panther, 1891 (first published Leipzig: Otto Wigand, 1845).
Englander, D. *A Documentary History of Jewish Immigrants in Britain 1840–1920*. Leicester: Leicester University Press, 1994.
Englander, D., and R. O'Day, eds. *Retrieved Riches: Social Investigation in Britain 1840-1914*. Aldershot: Ashgate, 2003 (first published 1998).
Evans, R. 'Rookeries and Model Dwellings: English Housing Reform and the Moralities of Private Space.' In *Translations from Drawing to Building and Other Essays*, edited by R. Evans, 92–117. London: AA Documents 2, 1997 (first published 1978).
Evans, R. *The Projective Cast: Architecture and Its Three Geometries*. London: Architectural Association, 1995.
Farr, W. 'Report on the Cholera Epidemic of 1866 in England: Supplement to the Twenty-Ninth Annual Report of the Registrar-General of Births, Deaths, and Marriages in England.' London: HMSO, 1868.
Farwell, W.B. *The Chinese at Home and Abroad Together with the Report of the Special Committee of the Board of Supervisors of San Francisco on the Condition of the Chinese Quarter of That City*. San Francisco: A. L. Bancroft & Co., 1885.
Fishman, W.J. *East End 1888: A Year in a London Borough among the Labouring Poor*. London: Gerald Duckworth & Co. Ltd., 1988.
Freedman, J. *Immigration and Insecurity in France*. Abingdon, Oxon; New York, NY: Taylor & Francis, 2017.
Frerichs, R.R. *Deadly River: Cholera and Cover-up in Post-Earthquake Haiti*. Ithaca: Cornell University Press, 2016.
Frerichs, R.R., P.S. Keim, R. Barrais and R. Piarroux. 'Nepalese Origin of Cholera Epidemic in Haiti.' *Clinical Microbiology and Infection* 18, no. 6 (2012): E158–E163.
Friendly, M. 'The Golden Age of Statistical Graphics.' *Statistical Sciences* 23, no. 4 (2008/11 2008): 502–35.
Friendly, M., and D.J. Denis. 'Milestones in the History of Thematic Cartography, Statistical Graphics, and Data Visualization.' http://datavis.ca/milestones/.
Gans, H. 'Planning and Social Life: Friendship and Neighbour Relations in Suburban Communities.' *Journal of the American Institute of Planners* 27 (1966): 134–40.
Gans, H. *The Levittowners*. New York: Pantheon Books, 1967.
Gans, H. 'Some Problems of and Futures for Urban Sociology: Toward a Sociology of Settlements.' *City & Community* 8, no. 3 (2009): 211–19.
Geddes, I., J. Allen, M. Allen and L. Morrisey. 'The Marmot Review: Implications for Spatial Planning.' London: Institute of Health Equity, UCL, 2011.
Geremek, B., and J. Birrell. *The Margins of Society in Late Medieval Paris*. Cambridge: Cambridge University Press, 2006.

Gieryn, T.F. 'City as Truth-Spot: Laboratories and Field-Sites in Urban Studies.' *Social Studies of Science* 36, no. 1 (2006, February 1): 5–38.

Gilbert, E.W. 'Pioneer Maps of Health and Disease in England.' *The Geographical Journal* 124, no. 2 (1958): 172–83.

Gilbert, P.K. 'The Victorian Social Body and Urban Cartography.' In *Imagined Londons*, edited by P.K. Gilbert. Albany: State University of New York Press, 2002.

Gillie, A. 'The Origin of the Poverty Line.' *Economic History Review* 49, no. 4 (1996, November): 715–30.

Ginn, G. 'Answering the "Bitter Cry": Urban Description and Social Reform in the Late-Victorian East End.' *The London Journal* 31, no. 2 (2006): 179–200.

Graham, R. 'New York City and Its Masters.' Edited by the Church Temperance Society of New York. New York: Church Temperance Society's Offices, 1887.

Great Britain. Parliament. House of Commons. 'The First Report of the Commissioners of His Majesty's Woods, Forests, and Land Revenues [Electronic Resource]: In Obedience to the Acts of 34 George III. Cap.75. and 50 George III. Cap. 65.' 494. London, 1812.

Griffiths, S. 'To Go with the Flows or to Flow with the Nodes? An Exploration of "Post-Disciplinary" Theories of Movement in Space Syntax and Mobilities Research.' In *11th International Space Syntax Symposium*, edited by T. Heitor (Chair), M. Serra, J.P. Silva, A. Tomé, M. B. Carreira, L.C. Da Silva and E. Bazaraite, 64.1–64.10. Lisbon, Portugal: University of Lisbon, 2017.

Griffiths, S., and L. Vaughan. 'Mapping Spatial Cultures: The Contribution of Space Syntax to Research in Social and Economic Urban History.' Paper presented at the Meeting of the European Association of Urban Historians. Helsinki, 24–27 August 2016.

Haine, W.S.C. *The World of the Paris Cafe: Sociability among the French Working Class, 1789-1914*. Baltimore, MD, and London: Johns Hopkins University Press, 1996.

Hanson, J. 'Urban Transformations: A History of Design Ideas.' *Urban Design International* 5 (2000): 97–122.

Hanson, J., and B. Hillier. 'The Architecture of Community: Some New Proposals on the Social Consequences of Architectural and Planning Decisions.' *Architecture et Comportement/ Architecture and Behaviour* 3, no. 3 (1987): 251–73.

Harrison, B. 'Pubs.' In *The Victorian City: Images and Realities (Past and Present & Numbers of People)*, edited by H.J. Dyos and M. Wolff, 161–90. London, Henley and Boston: Routledge & Kegan Paul, 1976.

Hebbert, M. 'Figure-Ground: History and Practice of a Planning Technique.' *Town Planning Review* 87, no. 6 (2016): 705–28.

Hellis, E.-C. *Memories of Cholera in 1832* (in French). Paris: Ballière, 1833. Source: Bibliothèque nationale de France, Department of Science and Technology, 8-TD57-389: http://catalogue.bnf.fr/ark:/12148/cb30589490z; Origin: National Library of France.

Hildyard, N. 'Building for Collapse.' *Ecologist* 7, no. 2 (Mar 1977): 46–54.

Hillier, A. 'Searching for Red Lines: Spatial Analysis of Lending Patterns in Philadelphia, 1940–1960.' *Pennsylvania History: A Journal of Mid-Atlantic Studies* 72, no. 1 (2005): 25–47.

Hillier, A. 'Invitation to Mapping: How GIS Can Facilitate New Discoveries in Urban and Planning History.' *Journal of Planning History* 9, no. 2 (2010): 122–34.

Hillier, B. 'In Defence of Space.' *RIBA Journal* (1973, November): 539–44.

Hillier, B. 'Can Architecture Cause Social Malaise?' In *Papers given to the MRC Conference on Housing*. London: Unit for Architectural Studies, UCL, 1991.

Hillier, B. 'Cities as Movement Economies.' *Urban Design International* 1, no. 1 (1996): 49–60.

Hillier, B. 'A Theory of the City as Object: Or, How Spatial Laws Mediate the Social Construction of Urban Space.' *Urban Design International* 3–4, no. 127 (2002).

Hillier, B. 'Credible Mechanisms or Spatial Determinism.' *Cities* 34 (2013): 75–7.

Hillier, B. 'The Common Language of Space: A Way of Looking at the Social, Economic and Environmental Functioning of Cities on a Common Basis.' *Journal of Environmental Science* 11, no. 3 (1998).

Hillier, B., R. Burdett, J. Peponis and A. Penn. 'Creating Life: Or, Does Architecture Determine Anything?' *Architecture & Behaviour* 3, no. 3 (special issue on space syntax) (1987): 233–50.

Hillier, B., and J. Hanson. *The Social Logic of Space*. 1990 ed. Cambridge: Cambridge University Press, 1990 (first published 1984).
Hillier, B., and L. Vaughan. 'The City as One Thing.' *Progress in Planning* 67, no. 3 (special issue on the syntax of segregation) (2007, April): 205–30.
Hobsbawm, E. 'Cities and Insurrections.' *Ekistics* 27, no. 162 (1969): 304–8.
Hubbard, P. 'Red-Light Districts and Toleration Zones: Geographies of Female Street Prostitution in England and Wales.' *Area* 29, no. 2 (1997): 129–40.
Huber, J.B. *Consumption, Its Relation to Man and His Civilization, Its Prevention and Cure*. Philadelphia: Lippincott, 1906.
Hume, A. *Condition of Liverpool, Religious and Social, Etc*. 2nd ed. Liverpool: Privately printed, 1858.
INEQ-CITIES. 'Paris: Socio-Economic Indicators.' UCL. Accessed 10 March 2018. https://www.ucl.ac.uk/silva/ineqcities/atlas/cities/paris/paris-sei#prim.
Jacobs, J. *The Death and Life of Great American Cities*. Harmondsworth, Middlesex: Penguin, 1961.
James, H. *The American Scene*. London: Chapman and Hall, 1907.
Jarcho, S. 'An Early Medicostatistical Map (Malgaigne, 1840).' *Bulletin of the New York Academy of Medicine* 50, no. 1 (1974): 96.
Johnson, S. *The Ghost Map*. London: Penguin Books, 2008.
Johnson, S.D., and K.J. Bowers. 'Permeability and Burglary Risk: Are Cul-De-Sacs Safer?'. *Journal of Quantitative Criminology* 26, no. 1 (1 March 2010): 89–111.
Johnston, R., J. Forrest and M. Poulsen. 'Are There Ethnic Enclaves/Ghettos in English Cities?'. *Urban Studies* 39, no. 4 (2002): 591–618.
Jones, T.J. 'Directory of Inhabited Alleys of Washington.' Washington: Housing Committee Monday Evening Club, 1912.
Karimi, K. 'The Spatial Logic of Organic Cities in Iran and the United Kingdom.' In *1st International Space Syntax Symposium*, edited by Major M. D., L. Amorim and F. Dufaux, 1–5 May 2017. London: University College London, 1997.
Katz, D. '"Clamber Not You up to the Casements": On Ghetto Views and Viewing.' *Jewish History* 24, no. 2 (2010): 127–53.
Katz, D.E. *The Jewish Ghetto and the Visual Imagination of Early Modern Venice*. New York: Cambridge University Press, 2017.
Keller, R.C. *Fatal Isolation: The Devastating Paris Heat Wave of 2003*. Chicago: University of Chicago Press, 2015.
Kershen, A. 'Henry Mayhew and Charles Booth: Men of Their Time.' In *Outsiders & Outcasts: Essays in Honour of William J. Fishman*, edited by G. Alderman and C. Holmes, 94–118. London: Duckworth, 1993.
Kershen, A. *Uniting the Tailors: Trade Unionism Amongst the Tailors of London and Leeds, 1870-1939*. Ilford, Essex: Frank Cass & Co., 1995.
Kershen, A. 'The Construction of Home in a Spitalfields Landscape.' *Landscape Research* 29, no. 3 (2004, July): 261–75.
Kershen, A., and L. Vaughan. 'There Was a Priest, a Rabbi and an Imam . . : An Analysis of Urban Space and Religious Practice in London's East End, 1685–2010.' *Material Religion* 9, no. 1 (2013): 10–35.
Khan, S.S., K. Karimi and L. Vaughan. 'The Tale of Ethno-Political and Spatial Claims in a Contested City: The Muhajir Community in Karachi.' In *Urban Geopolitics: Rethinking Planning in Contested Cities*, edited by J. Rokem and C. Boano. Routledge Studies in Urbanism and the City. Abingdon, UK: Routledge, 2017.
Kimball, M.A. 'London through Rose-Colored Graphics: Visual Rhetoric and Information Graphic Design in Charles Booth's Maps of London Poverty.' *Journal of Technical Writing and Communication* 36, no. 4 (2006): 353–81.
Klier, J.D. 'What Exactly Was a Shtetl?'. In *The Shtetl: Image and Reality*, edited by G. Estraikh and M. Krutikov, 23–35. Oxford: Legenda, published by the European Research Centre, 2000.
Klinenberg, E. *Heat Wave: A Social Autopsy of Disaster in Chicago*. Chicago: University of Chicago Press, 2005.
Kneale, J. 'The Place of Drink: Temperance and the Public, 1856–1914.' *Social & Cultural Geography* 2, no. 1 (2001): 43–59.

Kneale, J., and S. French. 'Mapping Alcohol: Health, Policy and the Geographies of Problem Drinking in Britain.' *Drugs: Education, Prevention and Policy* 15, no. 3 (2008): 233–49.

Koch, T. *Disease Maps: Epidemics on the Ground*. Chicago: University of Chicago Press, 2011.

Koch, T. *Cartographies of Disease: Maps, Mapping, and Medicine*. Redlands, CA: ESRI Press, 2017.

Koch, T., and K. Denike. 'Essential, Illustrative, or . . . Just Propaganda? Rethinking John Snow's Broad Street Map.' *Cartographica: The International Journal for Geographic Information and Geovisualization* 45, no. 1 (2010): 19–31.

Laurence, R., and D.J. Newsome. *Rome, Ostia, Pompeii: Movement and Space*. Oxford: Oxford University Press, 2011.

Law, S. 'Defining Street-Based Local Area and Measuring Its Effect on House Price Using a Hedonic Price Approach: The Case Study of Metropolitan London.' *Cities* 60 (2017): 166–79.

Lemon, A., and D. Clifford. 'Post-Apartheid Transition in a Small South African Town: Interracial Property Transfer in Margate, Kwazulu-Natal.' *Urban Studies* 42, no. 1 (2005): 7.

Linsley, C.A., and C.L. Linsley. 'Booth, Rowntree, and Llewelyn Smith: A Reassessment of Interwar Poverty.' *Economic History Review* XLVI, no. I (1993): 88–107.

Llewellyn Smith, H. 'The New Survey of London Life and Labour.' *Journal of the Royal Statistical Society* 92, no. 4 (1929): 530–58.

London, J. *The People of the Abyss (2014 Edition with Original Photographic Plates; Introduction by Iain Sinclair)*. London: Tangerine Press, 2014 (first published 1903).

Low Income Project Team for the Nutrition Task Force. 'Low Income, Food, Nutrition, and Health: Strategies for Improvement.' London: Department of Health, 1996.

Lowenfeld, J. 'Estate Regeneration in Practice: The Mozart Estate, Westminster, 1985–2004.' In *Twentieth Century Architecture 9: The Journal of the Twentieth Century Society – Housing the Twentieth Century Nation*, edited by E. Harwood and A. Powers, 163–74. London: The Twentieth Century Society, 2008.

Lupton, R. '"Neighbourhood Effects": Can We Measure Them and Does It Matter?' CASEpaper, 73. Centre for Analysis of Social Exclusion. London: London School of Economics, 2003.

Lynch, K. *The Image of the City*. Cambridge, MA: MIT Press, 1960.

MacCormac, R. 'An Anatomy of London.' *Built Environment* 22, no. 4 (1996): 306–11.

MacLeod, G., and K. Ward. 'Spaces of Utopia and Dystopia: Landscaping the Contemporary City.' *Geografiska Annaler: Series B, Human Geography* 84, no. 3–4 (2002): 153–70.

Major, M.D. '"Excavating" Pruitt-Igoe.' In *11th International Space Syntax Symposium*, edited by T. Heitor (Chair), M. Serra, J.P. Silva, A. Tomé, M. B. Carreira, L.C. Da Silva and E. Bazaraite. Lisbon, Portugal: University of Lisbon, 2017.

Marable, M. *W.E.B. Du Bois: Black Radical Democrat*. Boston: Twayne, 1986.

Marcuse, P. 'Putting Space in Its Place: Reassessing the Spatiality of the Ghetto and Advanced Marginality.' *City* 11, no. 3 (2007): 378–83.

Massey, D., and R. Denton. *American Apartheid: Segregation and the Making of the Underclass*. Cambridge, MA: Harvard University Press, 1993.

Mateos, P. 'Uncertain Segregation: The Challenge of Defining and Measuring Ethnicity in Segregation Studies.' *Built Environment* 37, no. 2 (2011): 226–38.

Matsipa, M. 'Street Values: Johannesburg's Spatial Agents.' *Architectural Review*, 4 July 2016.

Mayhew, H. *London Labour and the London Poor*. Vol. 4, London: Griffin, Bohn, 1861. Penguin Classics reprint edition, edited by V. Neuberg. Harmondsworth: Penguin, 1985.

Mayne, A. *The Imagined Slum: Newspaper Representation in Three Cities 1870–1914*. Leicester: Leicester University Press, 1993.

Mearns, A. 'The Bitter Cry of Outcast London: An Inquiry into the Condition of the Abject Poor.' London: James Clarke & Co., 1883.

Metzler, T. *Tales of Three Cities: Urban Jewish Cultures in London, Berlin, and Paris (1880-1940)*. Jüdische Kultur. Vol. 028, Wiesbaden: Harrassowitz, 2014.

Michelson, E. 'Conversionary Preaching and the Jews in Early Modern Rome.' *Past & Present* 235, no. 1 (2017): 68–104.

Michelson, E. 'Exiting the Roman Ghetto.' In *The Ghetto: From Venice to Chicago*, edited by D. Feldman, B. Cheyette and F. de Vivo. Birkbeck, University of London: Pears Institute for the Study of Antisemitism, 2017.

Mindell, J., P. Anciaes, A. Dhanani, J. Stockton, P. Jones, M. Haklay, N. Groce, S. Scholes and L. Vaughan. 'Using Triangulation to Assess a Suite of Tools to Measure Community Severance.' *Journal of Transport Geography* 60, no. April 2017 (2017): 119–29.

Mindell, J., and S. Karlsen. 'Community Severance and Health: What Do We Actually Know?' *Journal of Urban Health* (2012): 1–15.

Morris, A. *The Scholar Denied: W. E. B. Du Bois and the Birth of Modern Sociology*. Oakland, California: University of California Press, 2015.

Morrison, A. *A Child of the Jago*. Third edition. London: Methuen & Co., 1897.

Murtagh, B., and P. Shirlow. 'Spatial Segregation and Labour Market Processes in Belfast.' *Policy & Politics* 35, no. 3 (2007): 361–75.

Narvaez, L., A. Penn and S. Griffiths. 'The Spatial Dimensions of Trade: From the Geography of Uses to the Architecture of Local Economies.' *A|Z ITU Journal of Faculty of Architecture* 11, no. 2 (2015): 209–30.

Netto V. M., Pinheiro M. S. and R. Paschoalino. 'Segregated Networks in the City.' *International Journal of Urban and Regional Research* 39 (2015): 1084–102.

Newman, A. *Jewish East London: The Russell and Lewis Map (Publisher's Note on Reproduction of Map from 'The Jew in London')*. London: The London Museum of Jewish Life, 1985.

Nightingale, C.H. *Segregation: A Global History of Divided Cities*. Chicago: University of Chicago Press, 2012.

Noble, D., D. Smith, R. Mathur, J. Robson and T. Greenhalgh. 'Feasibility Study of Geospatial Mapping of Chronic Disease Risk to Inform Public Health Commissioning.' *BMJ Open* 2, no. 1 (2012).

Noorthouck, J. *A New History of London Including Westminster and Southwark*. London: privately printed, 1773. British History Online http://www.british-history.ac.uk/no-series/new-history-london.

Northall, J. *Travels through Italy: Containing New and Curious Observations on That Country*. London: Hooper, 1766.

O'Brien, J., and S. Griffiths. 'Relating Urban Morphologies to Movement Potentials over Time: A Diachronic Study with Space Syntax of Liverpool, UK.' In *11th International Space Syntax Symposium*, edited by T. Heitor (Chair), M. Serra, J.P. Silva, A. Tomé, M. B. Carreira, L.C. Da Silva and E. Bazaraite, 98.1–98.11. Lisbon, Portugal: University of Lisbon, 2017.

O'Brien, O. 'Geodemographics of Housing in Great Britain: A New Visualisation in the Style of Charles Booth.' http://vis.oobrien.com/booth/.

O'Day, R., and D. Englander. *Mr. Charles Booth's Inquiry: Life and Labour of the People in London Reconsidered*. London: Hambledon Press, 1993.

O.L.P. 'Letter to the Editor: An All-Night Saloon.' *New York Times*, 23 October 1910.

Orford, S., D. Dorling, R. Mitchell, M. Shawc and G. Davey-Smith. 'Life and Death of the People of London: A Historical GIS of Charles Booth's Inquiry.' *Health and Place* 8, no. 1 (GIS Special Issue) (March 2002): 25–35.

Osborne, T., and N. Rose. 'Spatial Phenomenotechnics: Making Space with Charles Booth and Patrick Geddes.' *Environment and Planning D: Society and Space* 22 (2004): 209–28.

Palaiologou, G. 'Between Buildings and Streets: A Study of the Micromorphology of the London Terrace and the Manhattan Row House 1880-2013.' Ph.D. diss., UCL, 2015.

Palaiologou, G., and L. Vaughan. 'The Sociability of the Street Interface – Revisiting West Village, Manhattan.' In *21st International Seminar on Urban Form - ISUF2014: Our*

Common Future in Urban Morphology, edited by V. Oliveira, P. Pinho, L. Batista, T. Patatas and C. Monteiro, 88–102. Porto, Portugal: FEUP, 2014.

Palsky, G. 'Connections and Exchanges in European Thematic Cartography. The Case of 19th Century Choropleth Maps.' *Belgeo. Revue Belge de Géographie*, no. 3–4 (2008): 413–26.

Parent-Duchâtelet, A.J.B. *On Prostitution in the City of Paris. From the French of M. Parent Duchâtelet*. Second edition. London: T. Burgess, 1840.

Park, R.E., and E. Burgess. *The City*. Chicago: University of Chicago Press, 1925.

Park, R.E. 'The City: Suggestions for the Investigation of Human Behavior in the City Environment.' *American Journal of Sociology* 20, no. 5 (1915): 577–612.

Park, R.E. 'The Urban Community as a Spatial Pattern and a Moral Order.' In *The Urban Community*, edited by E.W. Burgess. Chicago: University of Chicago Press, 1926.

Park, R.E., and E.W. Burgess. *Introduction to the Science of Sociology*. Chicago: University of Chicago Press, 1921.

Pattillo, M. 'Race, Class, and Crime in the Redevelopment of American Cities.' In *American Economies*, edited by E. Boesenberg, R. Isensee and M. Klepper, 143–59. Heidelberg: Universitatsverlag Winter, 2012.

Peach, C. 'The Ghetto and the Ethnic Enclave.' In *Desegregating the City: Ghettos, Enclaves and Inequality*, edited by D.P. Varady, 31–48. Albany: State University of New York Press, 2005.

Peach, C. 'Slippery Segregation: Discovering or Manufacturing Ghettos?' *Journal of Ethnic and Migration Studies* 35, no. 9 (2009): 1381–95.

Penn, A., and B. Croxford. 'Effects of Street Grid Configuration on Kerbside Concentrations of Vehicular Emissions.' In *1st International Space Syntax Symposium*, edited by Major M. D., L. Amorim and F. Dufaux, 27.1–27.10. London: University College London, 1997.

Perry, R. *Facts and Observations on the Sanitary State of Glasgow During the Last Year: With Statistical Tables of the Late Epidemic, Shewing the Connection Existing between Poverty, Disease, and Crime*. Glasgow: Glasgow Royal Asylum for Lunatics, printed at the institution, 1844.

Pfautz, H.W., ed. *On the City: Physical Pattern and Social Structure; Selected Writings of Charles Booth/ Heritage of Sociology*. Chicago: University of Chicago Press, 1967.

Philpott, T.L. *The Slum and the Ghetto: Neighborhood Deterioration and Middle-class Reform, Chicago, 1880–1930*. New York: Oxford University Press, 1978.

Pickering, W.S.F. 'Abraham Hume (1814–1884). A Forgotten Pioneer in Religious Sociology.' *Archives de sociologie des religions* 17, no. 33 (1972): 33–48.

Plunz, R. *A History of Housing in New York*. New York: Columbia University Press, 1990.

Polasky, J. 'Transplanting and Rooting Workers in London and Brussels: A Comparative History.' *The Journal of Modern History* 73, no. 3 (2001): 528–60.

Powell, R. 'Loïc Wacquant's "Ghetto" and Ethnic Minority Segregation in the UK: The Neglected Case of Gypsy-Travellers.' *International Journal of Urban and Regional Research* (2012): 115–34.

Princeton University Library. 'Medicine.' In *First X, Then Y, Now Z: Landmark Thematic Maps*. Princeton University Library Historic Maps Collection (2012). http://libweb5.princeton.edu/visual_materials/maps/websites/thematic-maps/quantitative/medicine/medicine.html. Accessed 11 April 2018.

Psarra, S. 'A Shapeless Hospital, a Floating Theatre and an Island with a Hill: Venice and Its Invisible Architecture.' In: *8th International Space Syntax Symposium*, edited by M. Greene (Chair), J. Reyes and A. Castro, 016:011–016:028. Santiago, Chile: Pontificia Universidad Católica de Chile (PUC), 2012.

Raban, J. *Soft City*. Glasgow: William Collins & Sons, 1974.

Ravid, B.C. 'From Geographical Realia to Historiographical Symbol: The Odyssey of the Word *Ghetto*.' In *Essential Papers on Jewish Culture in Renaissance and Baroque Italy*, edited by D.B. Ruderman, Essential Papers on Jewish Studies. New York and London: New York University Press, 1992.

Ravetz, A. *Council Housing and Culture*. London: Routledge, 2001.

Reeder, D. 'Charles Booth's Descriptive Map of London Poverty, 1889.' London: London Topographical Society, 1984.

Reeder, D. 'Representation of Metropolis: Descriptions of the Social Environment in *Life and Labour*.' Chap. 11 in *Retrieved Riches: Social Investigation in Britain 1840–1914* edited by D. Englander and R. O'Day, 323–38. Aldershot: Ashgate, 2003.

Residents of Hull-House – a Social Settlement. *Hull-House Maps and Papers: A Presentation of Nationalities and Wages in a Congested District of Chicago, Together with Comments and Essays on Problems Growing out of the Social Conditions.* New York: Thomas Cromwell, 1895.

Residents of Hull-House – a Social Settlement. *Hull-House Maps and Papers: A Presentation of Nationalities and Wages in a Congested District of Chicago, Together with Comments and Essays on Problems Growing out of the Social Conditions. Introduction by Rima Lunin Schultz. Includes Reproduction of Maps and Papers from 1895.* Urbana and Chicago: University of Illinois Press, 1895/2007.

Riis, J.A. *How the Other Half Lives: Studies among the Tenements of New York.* New York: Charles Scribner's Sons, 1890.

Rittel, H., and M. Webber. 'Dilemmas in a General Theory of Planning.' *Policy Sciences* 4, no. 2 (1973): 155–69.

Robinson, A.H. 'The 1837 Maps of Henry Drury Harness.' *The Geographical Journal* 121, no. 4 (1955): 440–50.

Robinson, A.H. *Early Thematic Mapping in the History of Cartography.* Chicago: University of Chicago Press, 1982.

Robinson, W.S. 'Ecological Correlations and the Behavior of Individuals.' *American Sociological Review* 15, no. 3 (1950): 351–7.

Robinson, W.S. *Muckraker: The Scandalous Life and Times of W.T. Stead: Britain's First Investigative Journalist.* London: Robson, 2012.

Roden, C. 'Food in London: The Post Colonial City.' Paper presented at the conference London: Post-Colonial City, Architectural Association, 12–13 March 1999.

Rokem, J., and L. Vaughan. 'Segregation, Mobility and Encounters in Jerusalem: The Role of Public Transport Infrastructure in Connecting the "Divided City".' *Urban Studies* (2017).

Rombai, L. 'Cartography in the Central Italian States from 1480 to 1680.' Chap. 36 in *The History of Cartography, Volume 3: Cartography in the European Renaissance, Part 1*, edited by D. Woodward, 909–39. Chicago: University of Chicago Press, 2007.

Roosevelt, T. *History as Literature.* New York: Charles Scribner's sons via Bartleby.com, 1998. http://www.bartleby.com/56/. 1913.

Ross, E. 'Survival Networks: Women's Neighbourhood Sharing in London before World War I.' *History Workshop*, no. 15 (1983): 4–27.

Rothenburg, J.N.C. *Die Cholera-Epidemie des Jahres 1832 in Hamburg: Ein Vortrag, gehalten im der wissenschaftlichen Versammlung des ärztlichen Vereins, am 17 November 1835.* Hamburg: Perthes & Besser, 1836.

Rowntree, B.S. *Poverty: A Study of Town Life.* Second edition. London: Macmillan, 1902 (first published 1901).

Russell, C., and H.S. Lewis. *The Jew in London (with a Map Specially Made for This Volume by Geo. E. Arkell).* London: Fisher Unwin, 1901.

Ryan, M. *Prostitution in London: With a Comparative View of That of Paris and New York, as Illustrative of the Capitals and Large Towns of All Countries: And Proving Moral Deprivation to Be the Most Fertile Source of Crime and of Personal and Social Misery: With an Account of the Nature and Treatment of the Various Diseases Caused by the Abuses of the Reproductive Function.* London: H. Bailliere, 1839.

Sampson, R.J. *Great American City: Chicago and the Enduring Neighborhood Effect.* Chicago: University of Chicago Press, 2012.

Sampson, R.J., and S. Raudenbush. 'Seeing Disorder: Neighborhood Stigma and the Social Construction of "Broken Windows".' *Social Psychology Quarterly* 67, no. 4 (2004): 319–42.

Samuel, R., ed. *East End Underworld: Chapters in the Life of Arthur Harding.* London: Routledge & Kegan Paul, 1981.

Sassen, S. 'The Global City: The De-Nationalizing of Time and Space.' Paper presented at the conference 'El territori en las societat de les xarxes. Dinámiques territorials i organització territorial', Barcelona, 2–3 October 2000.

Schierup, C.-U., and A. Ålund. 'The End of Swedish Exceptionalism? Citizenship, Neoliberalism and the Politics of Exclusion.' *Race & Class* 53 (2011), 45–64.

Schulten, S. *Mapping the Nation: History and Cartography in Nineteenth-Century America*. Chicago: University of Chicago Press, 2012.

Seamon, D. 'Lived Bodies, Place, and Phenomenology: Insights from Edmund Husserl, Maurice Merleau-Ponty, Edward Casey, Jane Jacobs, and Eric Klinenberg.' *Journal of Human Rights and the Environment* 4, no. 2 (special issue on human bodies and material space) (2013, September): 143–66.

Sennett, R. *Flesh and Stone – the Body and the City in Western Civilization*. New York: W.W. Norton and Company, 1994.

Sennett, R. *The Foreigner*. London: Notting Hill Editions, 2011.

Sennett, R. *The Uses of Disorder: Personal Identity and City Life*. London: Faber and Faber, 1996.

Shah, N. *Contagious Divides: Epidemics and Race in San Francisco's Chinatown*. University of California Press, 2001.

Shapiro, J. *A Space Syntax Analysis of Arroyo Hondo Pueblo, New Mexico: Community Formation in the Northern Rio Grande*. Arroyo Hondo Archaeological Series. Vol. 9, Santa Fe, New Mexico: School of American Research Press, 2005.

Shapter, T. *The History of the Cholera in Exeter*. London: John Churchill, 1832.

Shaw, C.R., and H.D. McKay. *Juvenile Delinquency and Urban Areas: A Study of Rates of Delinquency in Relation to Differential Characteristics of Local Communities in American Cities*. Chicago: University of Chicago Press, 1942.

Shaw, C.R., and F.M. Zorbaugh, *Delinquency areas: A Study of the Geographic Distribution of School Truants, Juvenile Delinquents, and Adult Offenders in Chicago*. Chicago: University of Chicago Press, 1929.

Shuttleworth, S., and S. Frampton. 'Constructing Scientific Communities: Citizen Science.' *The Lancet* 385, no. 9987 (2015): 2568.

Sibley, D. *Geographies of Exclusion: Society and Difference in the West*. London; New York: Routledge, 1995.

Simey, T., and M. Simey. *Charles Booth, Social Scientist*. London: Oxford University Press, 1960.

Simmel, G. 'The Metropolis and Mental Life (Original German, 1903).' In *The Sociology of Georg Simmel: Translated, Edited and with an Introduction by Kurt H. Wolff*, edited by K.H. Wolff, 409–19. New York and London: Macmillan Publishing (A Free Press Paperback), 1950.

Simmel, G. 'The Stranger.' In *The Sociology of Georg Simmel*, edited by K.H. Wolff, 402–8. New York: Free Press, 1950. First published in German as 'Exkurs über den Fremden.' In *Soziologie: Untersuchungen über die Formen der Vergesellschaftung*. Leipzig: Duncker & Humblot, 1908.

Smith, D. *Victorian Maps of the British Isles*. London: Batsford, 1985.

Snow, J. *On the Mode of Communication of Cholera*. 2nd ed. London: John Churchill, 1855.

Soja, E.W. *Postmodern Geographies: Reassertion of Space in Critical Social Theory*. London: Verso, 1989.

Spicker, P. 'Poverty and Depressed Estates: A Critique of *Utopia on Trial*.' *Housing Studies* 2, no. 4 (1987): 283–92.

Spicker, P. 'Charles Booth: The Examination of Poverty.' *Social Policy & Administration* 24, no. 1 (1990): 21–38.

Spicker, P. 'Poor Areas and the "Ecological Fallacy".' *Radical Statistics* 76 (2001).

Spinks, C. 'A New Apartheid? Urban Spatiality, (Fear of) Crime, and Segregation in Cape Town, South Africa.' Development Studies Institute Working Paper Series No. 01-20, 1–42. London: London School of Economics, 2001.

Stead, W.T. *If Christ Came to Chicago – A Plea for the Union of All Who Love in the Service of All Who Suffer*. London: The Review of Reviews, 1894.

Stedman Jones, G. *Outcast London: A Study in the Relationship between Classes in Victorian Society*. Oxford: Peregrine Penguin Edition, 1984.

Summers, L., and S.D. Johnson. 'Does the Configuration of the Street Network Influence Where Outdoor Serious Violence Takes Place? Using Space Syntax to Test Crime Pattern Theory.' *Journal of Quantitative Criminology* 33, no. 2 (2017, June): 397–420.

Sutherland, E.H. 'Delinquency Areas: A Study of the Geographic Distribution of School Truants, Juvenile Delinquents, and Adult Offenders in Chicago. Clifford R. Shaw, Frederick M. Zorbaugh, Henry D. Mckay, Leonard S. Cottrell (Review).' *American Journal of Sociology* 36, no. 1 (1930, July): 139–40.

Swensen, S. 'Mapping Poverty in Agar Town: Economic Conditions Prior to the Development of St. Pancras Station in 1866.' 1–62. London: London School of Economics, 2006.

Tarkhanyan, L. 'Drug Crime and the Urban Mosaic: The Locational Choices of Drug Crime in Relation to High Streets, Bars, Schools and Hospitals.' In: *9th International Space Syntax Symposium*, edited by Y.O. Kim (Chair), H.T. Park and K.W. Seo, 101:1–101:13. Seoul, South Korea: Sejong University, 2013.

Tecle-Misghina, B. *Asmara – an Urban History: Rivista L'architettura Delle Città – Unesco Chair Series N. 1*. Rome: Edizioni Nuova Cultura, 2015.

The Economist. 'There Goes the Neighbourhood.' 4 May 2006.

Thomas, B. 'The New Survey of London Life and Labour.' *Economica* 3, no. 12 (1936): 461–75.

Thrasher, F.M., and J.F. Short. *The Gang: A Study of 1,313 Gangs in Chicago*. Chicago: University of Chicago Press, 1963 (first published 1927).

Topalov, C. 'The City as Terra Incognita: Charles Booth's Poverty Survey and the People of London, 1886–1891.' *Planning Perspectives* 8 (1993): 395–425.

Uduku, O., and G. Ben-Tovim. 'Social Infrastructure in Granby/Toxteth: A Contemporary Socio-Cultural and Historical Study of the Built Environment and Community in "L8".' 1–57. Liverpool: Race and Social Policy Unit, University of Liverpool, 1998.

University of Chicago. 'Back Matter.' *Social Service Review* 4, no. 4 (1930).

Ury, S. 'Common Grounds? On the Place and Role of Jewish Coffee Houses at the Turn of the Century.' Paper given at a conference on *Warsaw – the History of a Jewish Metropolis*, June 22–25, 2010, University College London.

Valverde, M. *Law and Order: Images, Meanings, Myths*. Abingdon and New Brunswick, NJ: Rutgers University Press, 2006.

Van Dyck, D., E. Cerin, T.L. Conway, I. De Bourdeaudhuij, N. Owen, J. Kerr, G. Cardon, et al. 'Perceived Neighborhood Environmental Attributes Associated with Adults' Transport-Related Walking and Cycling: Findings from the USA, Australia and Belgium.' *International Journal of Behavioral Nutrition and Physical Activity* 9, no. 70 (2012): 1–14.

Varga, J.J. *Hell's Kitchen and the Battle for Urban Space: Class Struggle and Progressive Reform in New York City, 1894–1914*. New York: Monthly Review, 2013.

Varner, D. 'Nineteenth Century Criminal Geography: W.E.B. Du Bois and the Pennsylvania Prison Society'. *Journal of Historical Geography* 59 (2018): 15–26.

Vaughan, L. 'The Unplanned "Ghetto": Immigrant Work Patterns in 19th Century Manchester.' Paper presented at 'Cities of Tomorrow', the 10th Conference of the International Planning History Society, Westminster University, July 2002.

Vaughan, L. 'The Relationship between Physical Segregation and Social Marginalisation in the Urban Environment.' *World Architecture* 185, special issue on space syntax (2005): 88–96.

Vaughan, L. 'The Spatial Form of Poverty in Charles Booth's London.' *Progress in Planning* 67, no. 3, special issue on the syntax of segregation (2007): 231–50.

Vaughan, L. 'The Spatial Syntax of Urban Segregation.' *Progress in Planning* 67, no. 3 (2007): 199–294.

Vaughan, L. 'The Ethnic Marketplace as Point of Transition.' Chapter 3 in *London the Promised Land Revisited*, edited by A. Kershen, 35–54. Studies in Migration and Diaspora. Farnham, Surrey: Ashgate, 2015.

Vaughan, L. 'The Paradox of City Walls: Enclosure, Boundary, Barrier.' *Lobby Magazine* 3, (2015): 100–3.

Vaughan, L., D.C. Clark, O. Sahbaz and M. Haklay. 'Space and Exclusion: Does Urban Morphology Play a Part in Social Deprivation?' *Area* 37, no. 4 (2005, December): 402–12.

Vaughan, L., and I. Geddes. 'Urban Form and Deprivation: A Contemporary Proxy for Charles Booth's Analysis of Poverty.' *Radical Statistics* 99 (2009): 46–73.

Vaughan, L., S. Griffiths and M. Haklay. 'Chapter 1: The Suburb and the City.' Chap. 1 in *Suburban Urbanities: Suburbs and the Life of the High Street* edited by L. Vaughan, 11–31. London: UCL Press, 2015.

Vaughan, L., and A. Penn. 'Jewish Immigrant Settlement Patterns in Manchester and Leeds 1881.' *Urban Studies* 43, no. 3 (2006, March): 653–71.

Vaughan, L., and K. Sailer. 'The Metropolitan Rhythm of Street Life: A Socio-Spatial Analysis of Synagogues and Churches in Nineteenth Century Whitechapel.' Chap. 9 in *An East End Legacy. Essays in Memory of William J Fishman*, edited by C. Holmes and A. Kershen, 184–206. London: Routledge, 2017.

Vitos, M., J. Altenbuchner, M. Stevens, G. Conquest, J. Lewis and M. Haklay. 'Supporting Collaboration with Non-Literate Forest Communities in the Congo-Basin.' In *CSCW '17: Proceedings of the 2017 ACM Conference on Computer Supported Cooperative Work and Social Computing*, pp. 1576–90. New York: ACM, 2017.

Volchenkov, D., and P. Blanchard. 'Ghetto of Venice: Access to the Target Node and the Random Target Access Time.' *Physics and Society* (2007), https://arxiv.org/abs/0710.3021.

Wacquant, L. *Urban Outcasts: A Comparative Sociology of Advanced Marginality*. Cambridge: Polity, 2007.

Wacquant, L. 'A Janus-Faced Institution of Ethnoracial Closure: A Sociological Specification of the Ghetto.' Chap. 1 in *The Ghetto: Contemporary Global Issues and Controversies*, edited by R. Hutchison and B. Haynes, 1–33. Boulder, CO: Westview Press, 2011.

Walford, E. *Hampstead: Rosslyn Hill*. Vol. 5: Old and New London, London: British History Online, 1878. http://www.british-history.ac.uk/old-new-london/vol5/pp483-494. Accessed 18 August 2017.

Walter, B. 'Irish/Jewish Diasporic Intersections in the East End of London: Paradoxes and Shared Locations.' In *La Place De L'autre*, edited by M. Prum, 53–67. Paris: L'Harmattan Press, 2010.

Walter, B. 'England People Very Nice: Multi-Generational Irish Identities in the Multi-Cultural East End.' *Socialist History Journal* 45 (2014): 78–102.

White, J. *Some Account of the Proposed Improvements of the Western Part of London: By the Formation of the Regent's Park, the New Street, the New Sewer, &C. &C. Illustrated by Plans, and Accompanied by Critical Observations*. Edited by J.E. Moxon. London: W. & P. Reynolds, 1814.

White, J. *Rothschild Buildings: Life in an East End Tenement Block 1887–1920*. London: Pimlico, 2003 (first published London: Routledge and Kegan Paul, 1980).

Williams, B. *The Making of Manchester Jewry 1740-1875*. Manchester: Manchester University Press, 1985.

Winskill, P.T. *The Temperance Movement and Its Workers: A Record of Social, Moral, Religious, and Political Progress*. London, Glasgow, Edinburgh and New York: Blackie, 1892.

Wirth, L. *The Ghetto (1988 Edition, with a New Introduction by Hasia R Diner)*. Studies in Ethnicity. Edited by R.H. Bayer. New Brunswick (USA) and London (UK): Transaction Publishers, 1928.

Wirth, L. 'Urbanism as a Way of Life.' *The American Journal of Sociology* 44, no. 1 (1938): 1–24.

Wood-Lamont, S. 'W.T. Stead's Books for the Bairns.' http://www.attackingthedevil.co.uk/worksabout/bairns.php.

Woods, R.A., ed. *The City Wilderness: A Settlement Study by Residents and Associates of the South End House Edited by Robert A. Woods, Head of the House, South End Boston*. Boston and New York: Houghton, Mifflin and Company, 1899.

Yelling, J.A. *Slums and Slum Clearance in Victorian London*. The London Research Series in Geography. Vol. 10. London: Allen and Unwin, 1986.

Zangwill, I. *Children of the Ghetto: A Study of a Peculiar People*. London: Heinemann, 1922 (first published 1892).

Index

Arkell, George, 8
    and the map of Jewish East London, 9, 146

back-to-backs *see* housing
blocks *see* urban blocks
Booth, Charles
    as empiricist, 67
    classification of poverty, 78
    conducting the survey, 69
    description of slum areas, 10
    legacy of his enquiry, 87
    mapping of London, 70–4
    on spatial impact of railway lines, 77, 80
    papers delivered to the Royal Statistical Society, 10, 69, 73, 74
    police interviews, 71, 73–4, 121, 149–1, 174, 178
    possible connection to Hume poverty maps, 67
Booth, William, 2, 59n.49
Boston, 18, 27, 143
Boundary Estate, 81
Brooklyn Daily Eagle (newspaper), 44
brothels, 104, 173–6 *see also* prostitution
Builder (newspaper), 170

cartography
    as a science, 7, 172
    ecological fallacy in, 8, 168, 171, 220
    golden age of, 6
    history of, 3–5, 8, 172, 218
    limitations of, 25, 37, 121, 218–20
    objectivity in, 4, 6, 8, 220
cartographic methods
    choropleth, 4, 31, 121, 172
    data collection, 139
    dot maps, 4, 27, 37, 50, 180
    planar, 4
    use of colour, 8–9, 11, 27, 35, 39 *see also* visual *under* rhetoric
chapels, 137
Chicago, 45, 51, 159 *see also* Hull-House
Chicago School, 92n.68, 107, 115, 129, 171, 192

Chinese
    Exclusion Act of 1882, 17
    in the US census, 162
    migration, 48
    tensions with indigenous population, 132
    see also San Francisco *see also under* settlement patterns
children
    and pubs, 99
    in poverty, 35, 44, 96, 105, 150, 169, 175
    injury from automobile accidents, 50–1
    in Arkell's study of Jewish East London, 146–7
    in Du Bois' study of Philadelphia, 107, 109, 113
    in Hull-House study of Chicago, 139
    in School Board Visitors' reports, 71, 118
    premature death, 94
    see also gangs
A Child of the Jago (book), 80
Children of the Ghetto (book), 167n.65
churches
    attendance, 62–3, 69, 87
    contributions to by the middle classes, 66
    contributions to by the poor, 96
    in Venice, 14
    importance as community institutions, 110
    mapped, 181, 198
    numbers in comparison with saloons, 186
    symbiotic relationship with pubs, 184
    visibility of, 152, 185
    see also temperance societies
city
    disorder as cause of urban problems, 25, 84
    move into from countryside, 16, 113–14, 129, 185
city blocks, 30–1, 34, 45, 88, 121, 131
    The 'Lung Block', New York, 42–4
city walls, 12–13, 171, 214

colour *see* use in cartography
commuting, 45, 50, 77, 184, 192 *see also* suburbanisation
concentric zones model of city growth, 192–4, 196
configuration, spatial, 7, 18–19, 34, 54, 61, 83, 123, 133, 147, 200
contagious diseases
  cholera
    and poverty, 63
    and theories of mode of transmission, 32, 35, 38
    due to lack of drainage or cleansing, 27, 29
    in contemporary times, 49, 218
    mapping, 25, 31, 37, 39, 52
    prevalence in low-lying areas, 26–7
    Snow, John and his study of, 35–8
    see also quarantine
  bubonic plague, 2, 17
  tuberculosis
    and the lung block, 42–3
    blamed on building design, 45–7
    in contemporary times, 218
  yellow fever, 4, 25, 48
co-presence, 210
crime
  and delinquency, 194
  associated with drunkenness and poverty in Philadelphia, 108–9
  associated with drunkenness in York, 94, 177–8
  associated with spatial segregation in Chicago, 196–8
  environmental criminology, 200–1
  on Booth map of London, 75
  on Chinatown map of San Francisco, 136
  on Hume map of Liverpool, 63–4, 66
  problems with mapping of, 168
  see also disorder
  see also prostitution

da Vinci, Leonardo, 3–4
darkness
  associated with slums, 2, 73
  restrictions on movement on the Jewish people of Venice, 13
  used graphically to suggest immorality or poverty, 8–10, 27, 63, 70
  used graphically to suggest disease, 35, 39
  see also visual *under* rhetoric
deprivation, 66, 85, 87, 131, 159, 205–6
  see also index of multiple deprivation
design
  regulation of housing, 30, 38, 40–1, 44–5, 88, 122
  of street layouts, 39, 50, 80–1, 84, 88, 115, 123 *see also* Haussmann *see also* projects
diet
  as an aspect of culture, 162
  of manual labourers, 94, 96
  related to disease, 207
dirt
  and disorder, 70, 161, 169, 174
  as 'matter out of place' in anthropological thought, 177
  associated with disease, 27, 33
  associated with prostitution, 177
  rubbish and refuse, 26, 29, 33, 68, 112, 161, 177, 181
disorderly conduct, 170–1, 174–5
disease *see* contagious disease *see also* non-contagious disease
diversity
  ethnic or religious, 6, 42, 155, 214
  land uses, 161, 225, 227
  social class, 105
  the role of cities in, 6, 76, 131
drunkenness, 69, 77, 108, 170, 179
  see also pubs
Du Bois, W.E.B.
  as empiricist, 107
  classification of poverty, 109, 113
  conducting the survey, 108
  definition of the 'colour-line', 155–6
  lack of independent funds, 108
  legacy of his enquiry, 107, 115
  mapping of Philadelphia, 110, 112
  spatial conception of segregation, 109
dwellings *see* housing

ecological fallacy *see* cartography
Economist (newspaper), 86
employment
  discrimination in recruitment, 110, 162
  irregularity, 68–9, 104, 208 *see also* unemployment
encounters, cross cultural, 18, 155, 205, 210–11, 216 *see also* co-presence
Engels, Friedrich, 16
environmental determinism, 198
ethnicity *see* race
exclusion, racial, 17, 107, 131, 151, 208
  see also ghetto *see also* quarantine

Fabian Society, 90n.21
fire insurance plans, 2, 164n.13
food *see* diet

gambling, 117, 133, 185
gangs, 170, 196–9
geographical information science (GIS), 112
ghetto
  of Frankfurt, 21n.27

of Rome, 12, 209
of Venice, 12–14, 209–10, 216–17
The Ghetto (book by Wirth), 129
see also under segregation
gin palaces, 178–80, 184 see also pubs
gypsies, 16, 119

Haussman, Baron Georges-Eugène, 15, 53, 84
Hill, Octavia, 69
housing
  and disease, 27, 29, 32, 35, 42, 50
  and immorality, 170
  and poverty, 71, 80–1, 101, 122, 146, 175
  back-to-backs, 30–1, 64
  benefits or social aid, 86
  density, 30, 146
  deterioration, 122–4, 146, 171, 191
  discrimination, 110, 157
  gated, 79
  social, 45, 116, 119
  street-facing, 104, 123
  tenements, 10, 40, 101–5, 119, 122, 141, 170
Hull-House, 99–106, 139–43, 161
  see also settlement houses
Hume, Reverend Abraham, see Liverpool

immorality
  associated with intemperance, 178
  associated with living conditions, 16, 63, 66, 169–70, 174
  associated with poverty, 45, 109
  contagion, 15, 135, 170, 171, 173–4
  spatial mechanisms of 198 *passim*
  use of darkness to convey, 8, 176
  see also crime
index of multiple deprivation, 122, 206
industry
  in Asmara, 154
  in Chicago 198
  in Leeds, 27, 29
  in Liverpool, 66
  in London, 70, 80, 116
  sweated industries, 70
integration
  social, 124, 130, 155, 159, 206, 208, 214
  spatial, 84, 97, 131, 148, 152, 162
interdependence, 76, 150
interaction, 52, 85, 125, 129, 150

James, Henry, 22n.45
Jerusalem, 6, 211
Jewish
  see Ghetto of Venice
  see also under settlement patterns

La Guardia, Fiorello, 122
labour see work
Labour, UK government, 115
Labour, US Department of, 139
land use
  mapped, 133, 163
  patterns, 29, 45, 55, 75, 121, 200
language
  as an aspect of culture, 49, 162
  moralistic, 73, 133, 177
  sensationalist, 80, 175–6 see also under rhetoric
light
  colour on maps, 76
  illumination of buildings, 179
  lacking in buildings, 40, 88, 101
  see also darkness
linguistic see language
Leeds, 27–9, 141
Liverpool, 62–7, 86, 177–8
London
  East End, 10, 70, 73, 83–7, 146–52, 161–2, 216
  West End, 77, 180–1
Los Angeles, 125
Los Angeles Herald (newspaper), 45

mapping see cartography
marginal separation (space syntax theory), 75, 105, 131
markets
  labour, 16, 19, 68, 71, 171, 206, 208
  property, 110, 206
  street, 40, 73, 97, 98, 149, 155, 216
  London's Covent Garden, 184
  London's Petticoat Lane, 161
Mayhew, Henry, 10, 168
Mearns, Reverend Mearns, 68, 175
Miami, 157–8
middle-classes, 10, 66–7, 84, 86, 158, 170, 215 see also classification of poverty under Booth, Charles
migration see under settlement patterns
morphology see design

Newark, 143–4
New York
  drink maps, 185–91
  Hell's Kitchen, 170
  housing in, 40–2
  poverty in, 11, 122
    see also Riis, Jacob
  yellow fever map, 4–5, 25–6
New York State Commission, 45
non-contagious diseases, 49
  hernia, 8
  obesity, 50, 55, 207
  tetanus, 48

overcrowding, 87–8 *see also* housing
Oxford, 27

Pall Mall Gazette (newspaper), 175
Parent-Duchâtelet, Alexandre Jean Baptiste, 4, 173
Paris, 4, 15, 52–3, 131 *see also* Haussman
pedestrians
  activity, 97–8, 123
  injuries, 50–1
permeability, 217
Philadelphia, 125, 158–9 *see also* Du Bois
policing, 10, 108, 170, 173
poverty line,
  defined by Booth, 80, 84, 86
  defined by Llewellyn Smith, 117–18
  defined by Rowntree, 94, 99
poverty persistence, 53, 85
projects (as housing solution), 122–3
prostitution
  in Chicago, 175
  in Chinatown, 133–7, 174
  in Paris, 4, 173
  in Liverpool, 178
  in London, 150, 169, 174, 185
  in New York, 188
  in the film *Taxi Driver*, 177
  location on the periphery, 12, 171
  spatial regulation, 173, 177
public health
  1842 Liverpool Sanitary Act, 64
  1848 Public Health Act, 37, 39
  1875 Public Health Act, 39
  inspectors, 68
  practitioners, 38, 49
  science (and epidemiology), 39, 35, 39, 55
  see also under cholera
public realm, 141, 161, 210
public houses *see* pubs
public transport *see under* transport
pubs
  in Rowntree's study of York, 96–9
  for prostitution, 99, 185
  maps, 97–9, 179–1
  meeting place for clubs or societies, 96, 98, 185
  refuge from overcrowded dwellings, 215
  source of information for work, 125n.5
  see also gin palaces

quarantine, 2, 17, 24

race, 161–2
racism, 3, 154, 159, 207
redlining, 144, 153–9, 215

religion
  see churches, synagogues, temples
rhetoric
  linguistic, 17, 132–3, 198
    see also language
  visual, 17, 103, 133, 136–7, 143, 150, 169, 175 *see also* darkness
Riis, Jacob, 11, 41, 281
Rouen, 31–3
Rowntree, Joseph, 181
Rowntree, Benjamin Seebohm
  as empiricist, 93
  classification of poverty, 94
  description of slum areas, 96–7
  map of pubs, 97–9
Ruskin, John, 69

San Francisco, Chinatown, *see under* settlement patterns
School Attendance Officers, 118
School Board Visitors, 71, 118, 146
scientific methods
  map boundaries, 38, 139, 175, 219
  sampling, 94, 115, 122, 139, 158, 219
  scale of analysis, 160
  see also cartography
  see also unit of analysis
segment maps, 226 *see also* space syntax
segregation
  as a spatial concept, 129–31, 139, 141, 162, 214
  racial, 109–10, 215
  social or economic, 81
  ethnic or religious, 142, 163, 211
  spatial, 13, 208–9
settlement houses, 74
  see also Hull-House
settlement patterns
  African, 212, 216
  African-American, 51, 131, 157, 215
    see also Du Bois
  Chinese, 132–8, 211
  Jewish, 8, 142–4, 191–3
slumming (as urban exploration), 10–11
social housing *see under* housing
space syntax analysis, 17–18 *see also* marginal separation
  of Leeds, 29, 34
  of Liverpool, 66
  of London, 83–6
  of York, 97
  of New York, 188–9
street morphology
  analogy to insects, 16
  enclosure, 53, 76, 77, 170 *see also under* Booth, Charles
  labyrinthine or maze-like street patterns, 81, 120, 132–3, 170
  opening up, 170
Street Scene (opera), 11

suburbanisation
    as retreat from the city, 64, 66, 115
    as solution to urban overcrowding, 45, 51, 80, 88, 160
    as stage in immigrant settlement, 192–3
synagogues, 137, 151–3, 181, 210

Taxi Driver (film), 177
temples, 136
tenements *see under* housing
Tenement House Committee, 41–5
temperance societies, 98, 178–92
    see also churches
topography, 66, 156
    interstitial areas, 34, 64, 112, 197–8
    low-lying areas, 26–7
    severance, 39–40, 77, 124, 198, 207
transport, public, 123, 211

unemployment, 50, 68–9, 159, 185
unit of analysis
    blocks, 141, 143
    lots or plots, 101, 110, 160
    street segments, 70, 143, 147

urban periphery 12, 53, 76, 115, 131, 184, 213

Venice *see under* ghetto
verticality, 53, 217
vice *see* crime
visual rhetoric *see under* rhetoric

walkability, 55
Washington, 45, 47
Webb, Beatrice *née* Potter, 69, 125n.11
work
    casualisation of, 94
    and distance from home, 80
    manual, 66, 94, 116, 132, 174
    skilled, 116
    see also markets

York *see* Rowntree, Benjamin Seebohm

zoning
    racial, 151, 154–5
    see also concentric zones model

www.ingramcontent.com/pod-product-compliance
Lightning Source LLC
LaVergne TN
LVHW052114130426
836100LV00001B/1